T0144090

Intelligent Copyright Protection for Images

Chapman & Hall/CRC Computational Intelligence and Its Applications

Series Editor
Siddhartha Bhattacharyya

Intelligent Copyright Protection for Images
Authored: Subhrajit Sinha Roy, Abhishek Basu, and Avik Chattopadhyay

Intelligent Copyright Protection for Images

Subhrajit Sinha Roy, Abhishek Basu,
and Avik Chattopadhyay

CRC Press
Taylor & Francis Group
Boca Raton London New York

CRC Press is an imprint of the
Taylor & Francis Group, an **informa** business
A CHAPMAN & HALL BOOK

CRC Press
Taylor & Francis Group
6000 Broken Sound Parkway NW, Suite 300
Boca Raton, FL 33487-2742

International Standard Book Number-13: 978-0-367-19817-6 (Hardback)

This book contains information obtained from authentic and highly regarded sources. Reasonable efforts have been made to publish reliable data and information, but the author and publisher cannot assume responsibility for the validity of all materials or the consequences of their use. The authors and publishers have attempted to trace the copyright holders of all material reproduced in this publication and apologize to copyright holders if permission to publish in this form has not been obtained. If any copyright material has not been acknowledged please write and let us know so we may rectify in any future reprint.

Visit the Taylor & Francis Web site at
http://www.taylorandfrancis.com

and the CRC Press Web site at
http://www.crcpress.com

Mr. Subhrajit Sinha Roy would like to dedicate this book to his parents, Mr. Kunal Kumar Sinha Roy and Mrs. Jolly Sinha Roy.

Dr. Abhishek Basu would like to dedicate this book to his parents, Mr. Amiya Kumar Basu and Mrs. Namita Basu, his wife, Mrs. Sangita Basu, and his daughter, Ms. Aadrita Basu.

Dr. Avik Chattopadhyay would like to dedicate this book to his parents, Mr. Debabrata Chattopadhyay and Mrs. Bhabani Chattopadhyay, and his wife, Mrs. Debasmita Chattopadhyay.

Contents

Preface

HIGH-QUALITY DATA AUGMENTATION IS required to deal with the rapid growth of the multimedia data market. The digital domain offers exceptional precision and speed in data editing and reproduction of the multimedia objects. But, at the same time, the accessibility of multimedia items, such as music, films, books, or software, in digital form makes them vulnerable to large-scale, unauthorized, or illegal attacks. Therefore, it is necessary to implant copyright in order to defeat these attacks. Scientists and researchers have developed several different methods for copyright protection in the digital domain. Digital watermarking is one of the most effective techniques for copyright protection for multimedia objects.

The aim of this book is to introduce a digital image watermarking technique that embeds copyright information (known as a watermark) into any digital image. The nature of the human visual system has been utilized in making the mark perceptually transparent. Intelligent techniques, such as visual saliency generation and image clustering algorithms, are employed to achieve improved data transparency while increasing the quantity of hidden information. Thus, this book explains how intelligent technique-based image watermarking schemes can prevail over conventional practices.

Chapter 1 introduces this book with a brief study of information hiding and copyright protection together with a description of different hiding techniques used in the past. This chapter also introduces the concept of digital watermarking being used as a copyright protection tool. An overview: the basic operations, the classification, and the properties of a good digital watermarking system with their different applications and impairments, are concisely discussed in this chapter.

Chapter 2 deals with digital image watermarking from different points of view, starting with digital images that are mainly used in the implementation and assessment of digital watermarking schemes. This chapter highlights various techniques related to spatial domain and frequency domain digital image watermarking. Additionally, some state-of-the-art frameworks on digital watermarking are also considered in this chapter.

Chapter 3 addresses intelligent systems used in the proposed copyright protection system. The scheme involves saliency as well as the K-means clustering algorithm to implement the proposed technique. This chapter looks at saliency and salient objects, the purpose of saliency in watermarking, saliency detection and saliency maps, and various saliency map algorithms, followed by intelligent image clustering, which utilizes the K-means algorithm.

Chapter 4 demonstrates how intelligent techniques have been utilized in a copyright protection framework to implement a digital image watermarking scheme. The technique is proposed as a solution for overcoming the trade-off between robustness, imperceptibility, and data capacity. This chapter looks at the step-wise processes for watermarking, embedding, and extracting in detail.

Chapter 5 presents the hardware implementation together with the necessary architecture for the proposed image watermarking scheme. For execution, field programmable gate array (FPGA) has been chosen because of its shorter design time and lower cost when compared with application-specific integrated circuit (ASIC). This chapter is an overview of hardware realization for digital watermarking, using FPGA architecture for the watermark insertion and extraction systems.

Chapter 6 analyzes the investigative outcomes of the proposed scheme described in Chapter 4. Payload capacity, imperceptibility, and robustness are the three major parameters that are required to judge the efficacy of any image watermarking scheme. In this chapter, the system performance has been assessed on these three parameters and compared with some existing state-of-the-art watermarking frameworks. Subsequently, the hardware system evaluation has also been carried out in terms of computation speed, complexity, and utilization of logic components. A detailed analysis of performance of the proposed scheme, when compared with other modern schemes, validates its superiority. Moreover, it confirms that the proposed scheme can be an effective tool for copyright protection.

Finally, Chapter 7 sums up the significant conclusions that can be reached from this work and also suggests the direction for future work. The book concludes with the list of references used.

In this way, this book develops the basic concepts of digital watermarking as a copyright protection tool along with a novel intelligent technique-based image watermarking approach. It is hoped that the readers will enjoy reading this book as well as finding it to be a good resource for their research into copyright protection.

Acknowledgments

THE AUTHORS GRATEFULLY ACKNOWLEDGE the Focus Series on Computational Intelligence and Applications with CRC Press, UK, Taylor & Francis Group, as well as Professor (Dr.) Siddhartha Bhattacharyya for giving them the opportunity and assistance to write this book.

The authors would like to express their deep sense of worship and indebtedness to their respective organizations and authorities, respectively. The authors would also like to express their appreciation to staff and colleagues at the Global Institute of Management and Technology, RCC Institute of Information Technology, and University of Calcutta for their encouragement, kindness, and friendly cooperation in providing the means to carry out this work. The authors are extremely grateful to their research group and collaborators; without their help, this book would not have seen daylight.

We would like to express our sincere gratitude to our friends and family members for their constant encouragement and support over the course of writing this book.

Needless to say, without all the above help and support, the writing and production of this book would not have been possible.

Mr. Subhrajit Sinha Roy
Electronics and Communication Engineering
Global Institute of Management and Technology

Dr. Abhishek Basu
Electronics and Communication Engineering
RCC Institute of Information Technology

Dr. Avik Chattopadhyay
Institute of Radio Physics and Electronics
University of Calcutta

About the Authors

Mr. Subhrajit Sinha Roy received his B.Tech. in Electronics and Communication Engineering from West Bengal University of Technology in 2012 and M.Tech. in Telecommunication from the RCC Institute of Information Technology (under West Bengal University of Technology), Kolkata, India, in 2015. At present, he is Assistant Professor in the Department of Electronics and Communication Engineering at Global Institute of Management and Technology, Krishnagar, West Bengal, India. He is also enrolled in the Department of Radio Physics and Electronics under Calcutta University for his Ph.D. in Engineering and Technology. To date, he has six research publications in international journals, conference proceedings, and edited volumes to his credit. His fields of interest span digital image processing, information hiding, FPGA-based system design, low-power VLSI design, and quantum computing.

Dr. Abhishek Basu received his B.Tech. in Electronics and Telecommunication Engineering from West Bengal University of Technology in 2005, his M.Tech. in VLSI Design from the Institute of Radio Physics and Electronics, University of Calcutta, India, in 2008, and his Ph.D. (Engg) from Jadavpur University, India, in 2015.

He is currently Assistant Professor and Head of the Department of Electronics and Communication Engineering at the RCC Institute of Information Technology, Kolkata, India. He served as Undergraduate Program Coordinator in the aforementioned department from March 30, 2016 to January 1, 2017. Prior to this, he was Lecturer of Electronics and Communication Engineering at the Guru Nanak Institute of Technology, Kolkata, India, from 2008 to 2009. Before that, he served as Lecturer of Electronics and Communication Engineering at the Mallabhum Institute of Technology, Bishnupur, India, from 2005 to 2007. He is co-editor of a book and has more than 40 research publications in international journals, conference proceedings, and edited volumes to his credit.

He has held the following positions: Technical Co-chair of the Third IEEE International Conference on Research in Computational Intelligence and Communication Networks, November 3–5, 2017; Co-organizing Secretary of First International Symposium on Signal and Image Processing, November 1–2, 2017; Program Chair of the National Level Conference on "Frontline Research in Computer, Communication and Device (FRCCD 2015)," December 29–30, 2015; member of the Organizing Committee and Technical Program Committee at the National Conference on "Recent Innovations in Computer Science & Communication Engineering (RICCE 2016)," July 7–8, 2016; Hospitality Chair of International Conference on Intelligent Control, Power, and Instrumentation (ICICPI 2016) October 21–23, 2016. He is also a member of various organizing and technical program committees for several national and international conferences.

His research interests include digital image processing, visual information hiding, IP protection techniques, FPGA-based system design, low-power VLSI design, and embedded system design. Dr. Basu is a life member of the Indian Association for Productivity, Quality and Reliability.

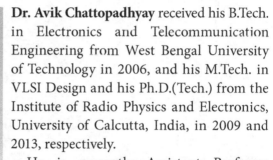 **Dr. Avik Chattopadhyay** received his B.Tech. in Electronics and Telecommunication Engineering from West Bengal University of Technology in 2006, and his M.Tech. in VLSI Design and his Ph.D.(Tech.) from the Institute of Radio Physics and Electronics, University of Calcutta, India, in 2009 and 2013, respectively.

He is currently Assistant Professor with the Institute of Radio Physics and Electronics, University of Calcutta, Kolkata, India. Prior to this, he served as Assistant Professor of Electrical and Electronics Engineering at the Birla Institute of Technology and Science, Pilani, Rajasthan, India, almost for two years. He is co-author of a book chapter, and has more than 20 research publications in international journals and conference proceedings to his credit.

His research interests include design and analysis of novel structures of MOSFETs, study of post-CMOS devices for low-power VLSI applications, and designing of FPGA-based systems.

CHAPTER 1

Introduction

1.1 INFORMATION HIDING AND COPYRIGHT PROTECTION

1.1.1 Overview

Information hiding is the study of communication in conceal-ment. The main purpose of this type of study is to prevent unau-thorized copying and to provide a secured or authenticated system for secret transmission. It may use covert channels, pseudocode writing, or embedding data into a cover object (Katzenbeisser and Petitcolas, 2000). Information hiding techniques are very useful in multilevel security systems. The need-based applications of information hiding can be categorized as follows:

- Information hiding may be used for automatic monitoring of the Web for copyright protection, i.e. to identify the poten-tial illegal usages of marked material on the Web or to tally the digestion of the downloaded documents of any registered Internet document against the original (Anderson, 1994).

- Information hiding is useful to indicate the particulars of music or advertisement content broadcasted on radio chan-nels (Willard, 1993). This is known as an automatic audit of radio transmission.

- For the benefit of the public, users, or purchasers, data augmentation is needed for any objects, such as music or images (Gerzon and Graven, 1995). Information hiding has a role in providing more efficient retrieval from databases of any hidden documents (Johnson, 1999).

- To detect any unauthorized modifications or illegal attempts on any document, a digital object can be embedded in the document so that any type of tampering can be detected (Lin and Chang, 1999).

- Authentication and royalty proof are matters for using information hiding technologies.

This book primarily focuses on the information hiding techniques that can carry out the function of copyright protection for multimedia objects. Basically, copyright is the intellectual information used to protect the uniqueness and ownership of original works or documents, such as images, text, audio, video, software, architecture, etc. The modern trend for digital data transmission leads to large volumes of information transfer. Customers are benefited by the ease of use of property in the digital domain. Regrettably, this feature is often mistakenly and poorly utilized and that degrades data authentication. So, copyright protection in data transmission is essential not least for providing authentication to the signals, and it has become one of the most vital issues in this modern research era.

In the next section, the progress of information hiding is drawn from the ancient age to the present day. This discussion will be helpful when developing ideas about the growth of information hiding techniques and their employment in copyright protection.

1.1.2 Hiding Techniques

According to the methods and purposes, we can classify information hiding techniques in several ways. Before we go into detail of the present-day methods, we look at a historical review of information hiding.

1.1.2.1 The Ancient Age

The practice of information hiding has been taking place since the age of verbal communication. When words were not invented and no languages were present, only a few different sounds served the purpose of conveying messages. The uniqueness was maintained by the sounds used by each community so that no one, other than the people of that community, could understand the messages. In the process of human evolution, we learned to write, and uniqueness was kept up in the writing patterns. Different languages evolved. Smaller groups amalgamated and used larger territories. People started to prefer uniformity in communication rather than individuality. At this stage came the need to hide secret information. There are several examples of covert writing and information hiding in the world's history. We can find them in Homer's *Iliad* (Homer, 1972), Herodotus' *Histories* (Herodotus, 1992), Vatsyayana's *Kama Sutra*, Kautilya's *Arthashastra*, and in many others (Ganguly, 1979; Kahn, 1967; Petitcolas et al., 1999). In the *Kama Sutra*, Vatsyayana mentions 'Mlecchita-vikalpa,' an art of secret writing through which women could conceal details of their liaisons. An example is given in Figure 1.1a of one of the suggested techniques. Here, letters from A to Z are haphazardly paired and the original message is written by replacing each letter

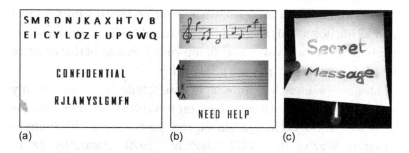

(a)   (b)   (c)

FIGURE 1.1 Example of early information hiding techniques: (a) Vatsyayana's cipher-based data hiding; (b) use of musical notation to hide information; (c) use of lemon juice as invisible ink.

with its comparable letter. Thus, the word CONFIDENTIAL has become RJLAMYSLGMFN. This led to present-day cryptography.

One of the oldest practices of information hiding is technical steganography, although the term 'steganography' itself came into use after the appearance of Trithemius' book on steganography in the fifteenth century. A famous example of technical steganography was told by Herodotus. In Ancient Greece, Histiaeus tattooed information on the shaved head of his most trusted slaves and, when the hair had regrown, the messages were hidden. Herodotus also told of how Demaratus, a Greek, informed Sparta about an imminent invasion by Xerxes, king of Persia. Removing the wax from a writing tablet, Demaratus wrote his message on the wood and covered it again with the wax.

Aeneas the Tactician (Tacticus, 1990) introduced some message-hiding techniques, such as sending letters through the soles of messengers' shoes, concealing messages in women's earrings, writing secret information on the wooden frames of writing tablets and then covering it them with wax, etc. He also suggested a famous technique that was still in use during the seventeenth century. Using this technique, a cover text was typed and the letters constructing the information were highlighted either by changing the heights of those particular letter strokes or by making very small holes above or below the letters (Katzenbeisser and Petitcolas, 2000). This technique was improved by Wilkins (1694), who used invisible ink when printing the tiny dots instead of making holes. This method was also used by German spies during both the World Wars, and an improved version of this practice is still in use for security purposes today.

Dragon, a French photographer, succeeded in making tiny images and, as a result, messages on microfilm became a means of information exchange via pigeon post during the Franco-Prussian War of 1870–1871 (Hayhurst, 1970; Tissandier, 1874). During the Russo-Japanese War (1904–1905), microscopic messages were hidden in ears and nostrils, and under fingernails (Stevens, 1968).

Linguistic steganography was also used for the purpose of information hiding. It was used in two ways: semagram, which is not in written form, and open code, which means a code that uses words in scrambled messages or illusions. As an example of a semagram, John Wilkins had shown how two musicians could communicate with each other by playing on their musical instruments as if they were actually speaking with those instruments. Musical notation can also be used as a secret language simply by identifying the staves with distinct alphabets (Figure 1.1b). Geometric drawings can use points, lines, angles, triangles, etc., which can also indicate a message. One of the greatest examples of open code is found in *Hypnerotomachia Poliphili* (Colonna, 1499), published in 1499. This book divulged an illicit love between a monk and a woman in an innovative way – the first letter of each of the 38 chapters spelled out *"Poliam frater Franciscus Columna Peramavil,"* which is supposed to mean "Father Francisco Columna loves Polia."

An extensively used technique of information hiding is the use of invisible inks. Lemon juice can be used as an ink, and it becomes visible only when the paper is heated (Figure 1.1c). Ovid suggested using milk as invisible ink in his *Art of Love*. Later progress in chemistry invented more sophisticated combinations of ink, developed at the time of the First World War. But the process was not so useful after the invention of 'universal developers,' which could be detected as the wetted parts of the paper. This led to the discovery of watermarking for information hiding.

An example of copyright protection from the ancient age is the *Liber Veritatis* of Claude Lorrain, the great painter. *Liber Veritatis* is a sketchbook that consists of two types of paper arranged in such a manner that every four blue pages will be followed by four white pages. This book was a collection of Lorrain's 195 drawings and was made to prevent forgery, as well as to serve the purpose of authentication (Samuelson, 1995).

The concepts of earlier data hiding techniques have been competently improvised and utilized in modern data communication, where information is mostly conveyed in digital form. In the following section, different types of modern data hiding technologies are briefly discussed and then we switch this discussion to digital watermarking as a copyright protection tool.

1.1.2.2 The Modern Age

Data transmission methods have been revolutionized with the rapid progress of civilization. The primordial information has taken the form of multimedia objects as the digital domain is mostly preferred for data communication. Naturally, radical changes happened regarding information hiding over the last a few decades (Anderson and Petitcolas, 1998). The digital domain offers ease in data editing and in data augmentation, which can lead to unauthorized copying, one of the most common problems experienced by the owners. In this search, both data authentication and copyright protection have become challenging issues. Moreover, wireless data transmission in an unsecured environment requires secret transmission due to jammers or hackers. As a result, the practice of information hiding methods is getting more and more important every day. Depending on the purpose, present-day information hiding techniques (Pfitzmann, 1996) are characteristically of the following types: anonymity, covert channel, cryptography, steganography, and watermarking.

Anonymity: Anonymity means being nameless or having a false name. It is used to hide the meta-content (i.e. the sender and the recipient) of a message. We will not explain this topic further, because our goal is to discuss copyright-providing technology.

Covert Channel: A covert channel is hidden from the access control mechanism used by the operating system and cannot be detected or controlled by the hardware-based security mechanism.

Covert channels can be used for a single system as well as for a network. There are two types: storage channels and timing channels, with storage channels more commonly used than the latter. The main drawbacks of this information hiding scheme are the low signal-to-noise ratio, low data rates, and the ease of detection simply by monitoring the system performance.

Cryptography: The word 'cryptography' has a Greek origin that means 'secret writing.' In cryptography, the original message, known as plaintext, is encrypted into ciphertext. This ciphertext is completely different from the plaintext and no one but the receiver can decode the ciphertext to retrieve the original message. Cryptographic information conveyance is also an age-old art. For example, in the *Mahabharata*, the great Indian epic, some sort of covert communication (Ganguly, 1979) occurred when Vidura warned the Pandavas about Duryodhana's conspiracy to put the Pandavas on a fire while still alive. Vidura said this in the presence of everybody, including the Kauravas. Only Yudhishthira realized the significance of what Vidura said – no one else imagined that it was a warning. Another example of cryptography, Vatsyayana's secret writing, is illustrated in Figure 1.1a.

In modern cryptography, ciphers and keys carry the central role in encoding and decoding (Barr, 2002; Mao, 2004). Ciphers refer to the encryption and decryption algorithms. Each sender–receiver pair necessitates a unique cipher for secure communication. A key is a number (or a set of numbers) through which the cipher operates. If the same key (or keys) is used in encryption and decryption, this is called symmetric key cryptography. Otherwise, this is known as asymmetric key cryptography. Two types of asymmetric keys are found in cryptography – private and public. The private key is kept only by the receiver and used in one-to-one data transmission. In one-to-many communication, the public key is utilized and is declared to the users.

A standard block diagram corresponding to the cryptography scheme is shown in Figure 1.2. Here we find that the encryption algorithm generates the ciphertext from plaintext using an encrypting key. At the receiving end, the decryption algorithm uses a decryption key to decode the ciphertext, recovering the plaintext in its original form.

Steganography: The word 'steganography' comes from the Greek words 'stegano' meaning covered or protected and 'graphia' meaning writing (Kahn, 1996). It is the art or practice of

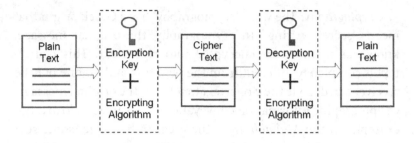

FIGURE 1.2   General block diagram for cryptographic data hiding.

concealing secret information in the form of a message, image, or file within another message, image, or object.

A good number of technical and linguistic steganographic examples are given in Section 1.1.2.1. Now we come to the more recent practice of steganography, digital steganography. Electronic communication can include steganographic coding inside a transport layer protocol, and this technique is called digital steganography (Anderson and Petitcolas, 1998). In digital steganography, information in the form of any multimedia object is hidden in another multimedia object through the embedding function, and a 'stego-document' is created. Only the receiver can retrieve the hidden data from the stego-document through the appropriate extraction algorithm. The process of retrieving information from a 'stego-object' is known as 'steganalysis.' The basic steganography technique is shown in Figure 1.3.

Digital Watermarking: Digital watermarking is the art of concealing information within any multimedia object in such a way that it can be extracted to verify the authenticity of the cover object (Bhattacharya, 2014; Mohanty, 1999).

Technically, digital watermarking is similar to digital steganography. But, the main differences between these two methods are that in digital watermarking the hidden information or watermark should carry novelty, copyright, or other relevant information regarding the cover; whereas, in case of steganography, the hidden data may have no relation to the cover. A detailed study on digital watermarking is covered in the next section.

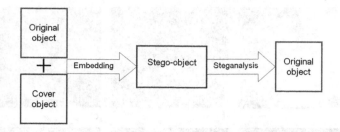

FIGURE 1.3   General block diagram for steganographic data hiding.

In this day and age, cryptography, steganography, and digital watermarking are mostly employed as information hiding techniques. But, as we have seen, they have some distinctions in their properties and application. To clarify this, a comparative study of these three information hiding techniques is demonstrated through the following simple example.

The image shown in Figure 1.4a requires secret transmission. (1) If the method chosen is cryptography, the original image is completely encrypted into a ciphertext, as shown in Figure 1.4b. Only by using the correct ciphers and keys can the original image be recovered from this encoded image (Figure 1.4c). (2) In the case of steganography, another image, completely different than the original image, is chosen as a cover, see Figure 1.4d. The secret image is imperceptibly implanted within the cover image and a stego-image is created (Figure 1.4e), with the original image remaining undetectable, except to the appropriate recipient. The correct steganalysis tool can recover the hidden information, as shown in Figure 1.4f. (3) Digital watermarking is employed when the issue is not to send the image in a covert way but to protect its originality or copyright. A unique mark, termed a watermark, can serve this purpose when it is embedded into the original image. End users can extract the watermark from the watermarked image and verify it against the original watermark to judge the authenticity of the received or watermarked image. The watermark, watermarked image, and extracted watermark are shown in Figures 1.4g–i respectively.

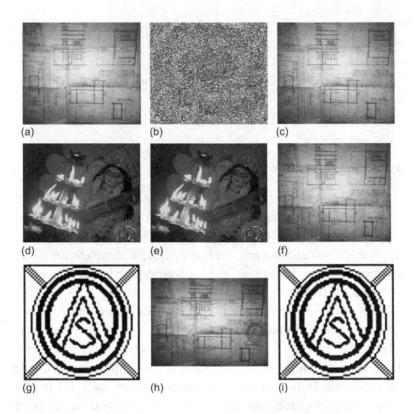

FIGURE 1.4 (a) Original image; (b) Encrypted image; (c) Decrypted image; (d) Cover image; (e) Stego-image; (f) Recovered image; (g) Watermark; (h) Watermarked image; (i) Extracted watermark.

The basic differences between cryptography, steganography, and digital watermarking can be summarized as shown in Table 1.1.

From this table, we can say that in digital watermarking the secret message is completely relevant to the cover. Later in the process, we can see that the message or watermark can be extracted from the cover without affecting the cover at all and it will still be present in the cover even after extraction. So, repeated authenticity verification can also be performed when using digital watermarking. Thus, digital watermarking has become the most

TABLE 1.1   Cryptography vs. Steganography vs. Watermarking

Cryptography	Steganography	Watermarking
The message itself converted into a distinct and unreadable form.	The message is made undetectable in a cover signal.	The message is made robust as the main priority.
The message needs a secret transmission.	Secret transmission is not required.	Secret transmission is not required.
Data quantity does not matter.	Data quantity should be controlled as it decreases the imperceptibility.	Data quantity should be controlled as it decreases the robustness.
It hides the content of the message.	It conceals the existence of the message.	It prevents the cover document from any types of unauthorized or illegal attempts.
The message is transmitted and, being encrypted, cover is not required.	The message may have nothing to do with the cover.	It carries the authenticity, information, or copyright of the cover object.
It is for point-to-point communication purposes.	It is for point-to-point communication purposes.	It is for one-to-many schemes.

preferred and most suitable approach for the purpose of copyright protection.

1.2 DIGITAL WATERMARKING AS A COPYRIGHT PROTECTION TOOL

1.2.1 Overview

Digital watermarking is the method of inserting information, called a watermark, into a multimedia object so that the watermark can be extracted or detected at the receiving end. The purpose of digital watermarking is to make an assertion about the object in which the watermark is embedded. In other words, it can be said that the watermark, embedded into any object, carries copyright for that particular object.

In fact, *paper* watermarking is an age-old art. A short time after the invention of paper, watermarking techniques began to be used during the paper manufacturing procedure (Berghel, 1997). At that time, the basic purpose of watermarking was to provide the maker's details. The reason it is called 'watermarking' is because of the process used. A semi-liquid blend of fiber and water was poured into a mesh and spread over the frame to provide an appropriate shape. Finally, the paper was produced in the form of compressed fiber obtained after completely removing the water by using pressure on the mixture. Before the fibers were compressed and dehydrated, an image or text from a negative was used to make an impression on the paper, leaving an enduring mark. This mark, being imposed on the paper through vaporization, was known as a 'watermark.' There is also a second theory about how this name came about. According to this theory, the watermarking technique originated about 700 years ago in Fabriano, Italy. The reason is considered to be the stance of the mark, rather than how it is made, and the method of imprinting the mark was also dissimilar to the one outlined above. In this process, an article was labeled in an unchangeable way by making a fragment of the article a little thin. When the paper containing the article as well as the mark was placed against a strong light source, the thin part of the content could be detected and it looked like a watery area; hence the term 'watermark' came about.

Digital watermarking is also a noticeable or concealed mark implanted into any digital data form, i.e. a multimedia object such as text, image, audio, or video. But clearly the insertion process for this insubstantial digital mark must differ from the paper watermarking techniques described above. For covert watermarks, the extraction of the mark is another important consideration. Therefore, we need to look at the basic operations of present-day digital watermarking schemes before we move on to a heuristic study of them.

1.2.2 Basic Operations

Digital watermarking is key for getting to the root of many issues regarding data authentication and copyright protection in the domain of digital or multimedia data transmission. Therefore, a variety of watermarking schemes are used to resolve the different concerns. But, from a generic viewpoint, there are three major operations that should be present in all the watermarking algorithms. As discussed above, we know that watermarking processes are largely intended to embed copyright information into any multimedia object before it is transmitted, to recover the information at the receiving end, and to confirm whether the extracted mark is authentic or not. So, the basic operations may be classified as follows:

(i) Watermark insertion or embedding

(ii) Watermark extraction or retrieving

(iii) Authenticity checking.

Watermark Insertion: The purpose of an embedding system is to implant information as an authentication or copyright mark into a multimedia object. There are two important terms that will be used several times in our discussion. The first one is 'watermark,' the definitions and uses of which are already known to us: it refers to the information to be embedded and it may be in the form of any multimedia object, dependent on its purpose. The second one is 'cover' or 'host,' which refers to the object into which the watermark is to be inserted. Any multimedia object, such as text, image, audio, and video, can be used as the cover object. One thing that it is important to keep in mind is that it is not essential that the cover and the watermark are of same type, i.e. not only an image but also text can be used as a watermark for a cover image.

Let W be the watermark and C the original cover object. Now, W is implanted into C through an exact embedding function, generating the corresponding watermarked object E. The embedding function may use a different set of functions to serve different, distinct purposes and a number of watermarking schemes are described in the following chapters. In general, if we consider the set of insertion related functions as a single function f_E, then the watermarked object E can be defined as:

$$E = f_E(C, W) \tag{1.1}$$

The general block diagram of a basic watermark embedding system is shown in Figure 1.5.

Watermark Extraction: Put simply, the purpose of watermark extraction is to retrieve the watermark from a received object. An analogous extraction function (i.e. analogous to the embedding function) is required to recover the watermark. For blind watermarking, the extraction function is able to recover the watermark from the received object (assumed to be watermarked) only (i.e. the original cover object is not required). When the original cover object is also required, this is known as non-blind watermarking. The extraction function can also be a combination of different functions. But, as above, a solo function can also be considered to be a collection of several interacting functions. If the extraction function is noted as f_R and the received object is R, then the watermark, W_R, extracted from it, can be defined as:

$$W_R = f_x(R) \tag{1.2}$$

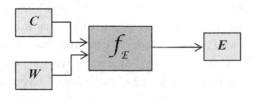

FIGURE 1.5 Generic block diagram of watermark embedding system.

or,

$$W_R = f_x(R,C) \tag{1.3}$$

Equation 1.2 is for blind extraction and Equation 1.3 is for non-blind extraction, considering C as the original cover object, to be taken as the reference input to the extraction system.

Authenticity Checking: Now, the fact is that, for any received signal, that signal may or may not consist of the exact desired watermark, but an extraction algorithm must generate an output. In this search, the needs of validity or authenticity test for the extracted watermark is considered. Here, W_R is first judged against the original watermark W through a similarity function, f_S, which is basically a set of qualitative and quantitative functions. The function's outputs are taken to be the inputs for a comparator. Some threshold values are defined as the maximum acceptance levels in favor of originality. The comparator analyzes the authenticity of the extracted watermark from the deviations between the outputs from f_S and their corresponding thresholds. The comparator sets its output as valid when the results are within the threshold range and as invalid otherwise. Figure 1.6 illustrates the simple block diagram for a standard extracting process along with authenticity checker.

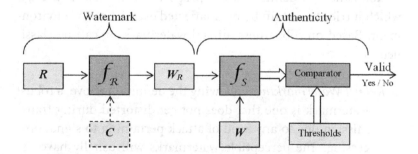

FIGURE 1.6 Generic block diagram for watermark extracting system with authenticity checker.

1.2.3 Classification

Digital watermarking can be classified in several ways. As explained above, any type of media object can be used as cover and, according to the nature of cover entity, watermarking can be categorized into four types. These are:

Text Watermarking (when text is chosen as the media object): Used for copyright protection in digital manuscripts, research papers, and e-books.

Image Watermarking (when an image is chosen as the cover): Used in digital cameras, image watermarking is used to implant legitimacy into photographs.

Audio Watermarking (when the cover should be an audio file): An artist can leave an assertion in his audio track through this type of watermarking.

Video Watermarking (when the cover object is a video): Frequently used in video broadcasting.

Consideration of the nature of a watermark is another means of classifying digital watermarking. There are mainly two characteristics of watermarks that can be categorized this way: the robustness and the perceptivity of the embedded information.

Robustness is defined as the property of an object through which it tries to sustain being unaffected even in 'noisy' environment. Based on robustness, digital watermarking can be classified as:

Robust Watermarking: Following the definition above, a robust watermark is one that does not get distorted during transmission due to any kind of attack pertaining to signal processing. The perceptible watermarks will usually have the characteristic of robustness.

Fragile Watermarking: In a contrast to a robust watermark, a fragile watermark can be manipulated easily by any attacks. It is, therefore, extremely useful for tamper detection.

Semi-Fragile Watermarking: If a watermark can resist some types of attack but is affected by others, it is known as a 'semi-fragile watermark.'

According to data perceptivity, a watermark is of two types:

Perceptible Watermarking: This type of watermark is easily detected through human perception (Figure 1.7a). It is used mainly when the credentials of the owner are required to be disclosed publicly.

Imperceptible Watermarking: These watermarks remain undetectable within the host object (Figure 1.7b). The undetectable nature of these watermarks enriches the aesthetic quality of the original cover document. This type of watermarking is employed for authenticity verification.

Imperceptible watermarking techniques are also categorized with respect to their extraction or detection process: blind

(a)   (b)

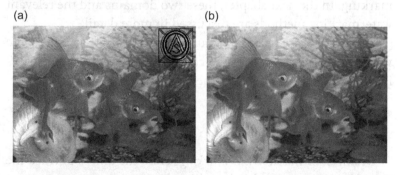

FIGURE 1.7 (a) Perceptible watermarking; (b) Imperceptible watermarking.

watermarking, non-blind watermarking, and semi-blind watermarking. If the original cover is required to retrieve the watermark from a received object, the extraction process is called a blind process. A non-blind technique never needs any reference to extract the mark from the received entity. For semi-blind watermarking, some additional information, but not the original cover, is required to extract the watermark.

After choosing the cover media and watermark properties, we need to look at the domain in which data insertion and extraction are to be performed. Based on the working domain, digital watermarking is of two types:

Spatial Domain Watermarking: In this practice, randomly or logically selected pixels are modified to embed the watermark.

Frequency Domain Watermarking: In this domain, watermarks are inserted into the suitable frequency regions of the host unit.

For spatial domain watermarking, computational cost and time are lower than those for frequency domain watermarking, whereas better robustness can be achieved with frequency domain watermarking. In the next chapter, these two domains and the relevant watermarking methods are discussed in more detail.

Finally, the application feature (i.e. the location where the watermark is inserted into the cover object) divides watermarking schemes into two classes. These are:

Source-Based Watermarking: Applications specified as ownership recognition are said to be 'source based.'

Destination-Based Watermarking: Here the aim is to trace the buyer in case of illegal reselling.

A simple example can be taken for a better understanding of this point. Suppose that when Jorge takes photographs, a copyright mark is implanted into each one. This is a source-based practice. When he comes to sell his photographs, he will sell them to many different people, including, for example, Pitter. However, Pitter makes a number of copies of Jorge's photographs, more cheaply than if purchasing them from Jorge. Jorge finds out about this, but he cannot do anything about it. So, going forward, Jorge puts another, different, technology into use. Along with the copyright mark, this technique embeds a watermark at the *receiving* end. Therefore, each and every buyer can be traced. This operation is known as a destination based method.

Figure 1.8 provides a simplified view of the classifications of digital watermarking from different aspects.

1.2.4 Properties for a Good Digital Watermarking Scheme

In this section, we will discuss some generic properties required for a good watermarking scheme.

Robustness: As discussed above, this is the property through which information, embedded in an object, remains unchanged during transmission. In other words, robustness is the property of a watermark to protect itself from digital signal processing activities, such as digital-to-analog conversion and vice versa, resampling, quantization, enhancement or lossy compression, and many others. At the same time, it also provides a resistance to general geometric distortion, like rescaling, cut, and rotation, as well as the operation, such as print, reprint, rescan, etc. Therefore, robustness is one of the most entailed properties of any digital watermarking scheme except it is fragile or semi-fragile in nature.

Imperceptibility: It means the quality of the cover will not be detracted in the aesthetic sense. Information embedded in an object should be imperceptible; meaning it will be neither visible nor hearable to human eyes or ears.

Data Capacity or Data Load: A watermark should try to increase the capability of embedded data quantity with a certainty of proper retrieval of the watermark in the extracting end.

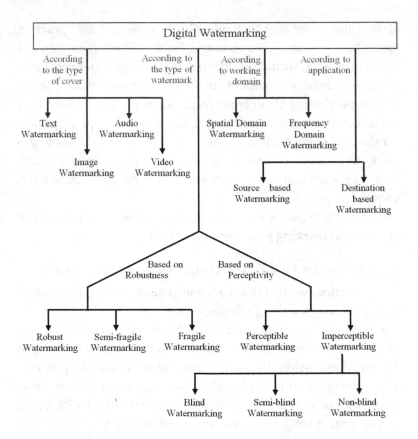

FIGURE 1.8   Classifications of digital watermarking.

Security and Reliability: Watermarks should deserve a unique sign so that the copyright protection purpose could be achieved.

Low Complexity: An algorithm should be made in such a manner that the scheme will be time effective, i.e. the watermark will be embedded and extracted using less time and effort.

This book mainly deals with the image watermarking scheme (Kutter, 1999). Now, with these general characteristics in mind, several types of image watermarking schemes have requirements of some definite features. Here are some specific features for different types of digital image watermarks (Mohanty, 1999):

- Desired Features for a Visible Watermark:

  - Secure Hiding Place: Watermarks should be embedded in an important or large area of the cover image so that destruction of the watermark by editing can be prevented.

  - Color Sensitivity: As the watermark is visible in nature, it should be sensitive for both monochromatic and color images.

  - Watermarks should never cause any effect that may make the original image blurred.

- Desired Features for an Invisible Robust Watermark:

  - The watermark should not be observed by the receiver.

  - It should not allow any type of degradation in quality.

  - It should carry a certain proof of the owner.

  - Even if the cover image is of high quality, modulation of watermark pixels should be minimized.

  - Most importantly, it cannot be removed, edited, or distorted. It should be resistive to most of the watermark attacks.

- Desired Features for Invisible Fragile Watermark:

  - It should have the property of invisibility.

  - It should be immediately modified in case of any type of pixel alternation in the host image.

  - It has to serve more security so that, in spite of being familiar with the scheme, one should not be able to regenerate the watermark after image alternation.

1.2.5 Application

Digital watermarking was introduced as one of the information hiding frameworks, primarily to serve the purpose of copyright protection. Day-by-day the application of digital watermarking spans a large area of new age digital communication and multimedia security. Some of the foremost applications are stated below:

Copyright Protection: The most noteworthy function of digital watermarking is definitely copyright protection. A large number of multimedia entities are always transmitted over networks. To meet extensive demand of these media objects, data augmentation is essentially carried out. But unauthenticated copying can also be performed at the same time. In this issue, copyright protection has become a crucial way out. As an example, anybody can use a number of images available on the Internet without any payment of royalty. Watermarks, acting as an ownership mark, can pin down the redistribution of the images.

Content Protection: When any content is embossed with a visible and robust watermark, it will signify the ownership or originality. In this way, contents can be distributed and made accessible through the Internet more freely and publicly. Library manuscripts often utilize watermarking for content protection.

Content Tagging: Content labeling is almost similar to content protection carrying some additional information about the object, for example, quality, manufacturer's description, etc. Here the purpose determines whether embedded information will be noticeable or not.

Authentication: In such applications, such as ID cards, credit cards, or ATM cards, the authenticity in controlling the cover entity needs to be confirmed. This purpose can be executed by implanting watermark data in addition to providing a confidential key to the card-holder to access the information.

Indication of Originality: We have discussed how copyright can put off the undermining of originators. Conversely, it is also a benefit to the procurer by leaving a mark as an evidence of ownership. The trader's watermark in the content object ensures that the entity belongs to the vendor and that it is not produced illicitly or without payment of royalties by copying or editing the object.

Embezzlement Detection: Misappropriation is a vital issue in digital retailing. It may occur that somebody has procured cost-generating items from a licensed proprietor and put these items on the market cheaply or free of charge, depriving the certified owner of the revenue. This kind of deceitful dealing can be controlled by using destination-based watermarking.

Labeling in Medical Reports: Nowadays, the digital watermarking is used in medical reports, like X-ray and magnetic resonance imaging (MRI). As the report is imperative for the patient, many problems may occur for any type of misplacement. Therefore, in these reports, digital watermarking can play a vital role.

Tamper Detection and Trustworthy Confirmation: One of the most competent uses of fragile watermarking is to detect any sort of tampering action performed on the host object. Here, if any alteration has happened, the watermark will be degenerated or distorted so that no modified version can be claimed to be original. In photography or capturing videos, it is used to indicate that the entity has been originally captured by a trustworthy camera (Friedman, 1993), not created by editing or fabricating any sight. Actually, at the time of capturing a picture or video, an invisible watermark is embedded into the object to avoid editing.

Digital Fingerprinting: Digital fingerprinting is practiced by using biometric information such as fingerprints as watermarks. It is utilized to discriminate an object from other comparable objects by offering distinctive fingerprints for every different object, as well as for each party. It can also be employed to perceive any alternation among the objects stored in a digital library. Some authors describe digital fingerprinting as a diverse methodology, not as a part of watermarking.

Broadcast Monitoring: One of the most significant aspects of watermarking is that it facilitates commercial groups to make sure classified advertisements, whether broadcasted through television or radio, appear for a precise period or not.

Source Tracking: To trace the information about the source history of any multimedia object delivered at the receiving end, digital watermarking is a functional tool.

1.2.6 Attacks

Attacks during signal processing are referred to as the deformations or modifications performed intentionally or accidentally on the transmitted objects. It is pretty difficult to recognize and categorize all signal processing attacks. The attacks identified up to now can be divided into five groups (Hernandez and Kutter, 2001):

Removal Attack: It aims to remove the embedded watermark and to decrease in the effective channel capacity.

Geometrical Attack: Here the intention is to distort the watermark rather than to remove it. The purpose is the same as the aim of resynchronization of a signal. It may also reduce the channel capacity or may fully destroy the watermark detection. This includes operations such as scaling, cropping, translation, rotation, etc.

Signal Processing Attack: This degrades the quality of data and includes digital-to-analog and analog-to-digital conversions, resampling, requantization, recompression, high- or low-pass filtering, nonlinear filtering, addition of noise (Gaussian and non-Gaussian), color reduction, etc.

Cryptographic Attack: The aim is to remove or destroy the embedded watermark by using a number of authorized recipients that can work together with differently watermarked copies, resulting in an unwatermarked copy, which uses a statistical average of all the watermarked images.

Protocol Attack: The purpose is to identify the weakness of a system label to prove a given watermarking method is not secure.

Some manual intentional attacks that can also be introduced are printing, rescanning, making forgeries (copying a watermarked image with a valid embedded watermark by unauthorized recipients), and IBM attack (to produce fake originals so that the fake owners could demand copyrights) (Borda, 2011). Moreover, we should keep in mind that some compression schemes, such as Joint Photographic Experts Group (JPEG) and Moving Picture Experts Group (MPEG), can cause a potential degradation in data quality.

CHAPTER 2

Perspectives on Digital Image Watermarking

2.1 INTRODUCTION TO DIGITAL IMAGE

The word 'image' comes from the Latin word *imago*, which stands for an artifact of depicting the external perception of any object, experienced through the human visual system (HVS). In other words, 'image' means the analogous portrayal of any visual stimuli in the form of a two-dimensional picture or photograph. In the case of digital image processing, an image, precisely termed as a 'digital image,' can be represented numerically through a two-dimensional array function $f(m,n)$, where m and n are the coordinates of the array plane with discrete finite values, and the function f also has distinct values (specifying the intensity of the image) within a finite range for all possible pairs of the coordinates (Gonzalez and Woods, 2008). This kind of depiction is often said to be the spatial representation of an image, see for example Figure 2.1. The zoomed portion clearly shows that the image is

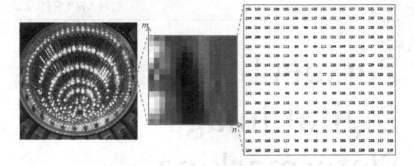

FIGURE 2.1 Two-dimensional array representation of image, composed of pixels with distinct values and locations.

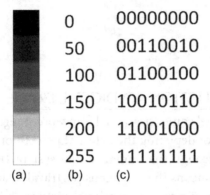

(a)	(b)	(c)
	0	00000000
	50	00110010
	100	01100100
	150	10010110
	200	11001000
	255	11111111

FIGURE 2.2 (a) Grayscale image pixels; (b) resultant intensity values; and (c) 8-bit representation of pixel values.

composed of a number of elements, known as pixels, having individual values and coaxial locations.

This type of image representation, for example as shown in Figure 2.1, is often known as a 256-level grayscale image having its pixel values within a range of 0 to 255. Each pixel is considered an 8-bit pixel. An increase in pixel values indicates a rise in intensity, as shown in Figure 2.2.

The bits from all the pixels together form several bit-planes. For an 8-bit pixel representation, there should be 8 bit-planes in that

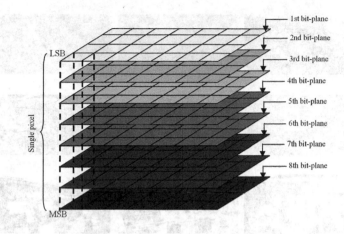

FIGURE 2.3   Bit-plane representation of an image.

particular image. The bit-plane concept is shown in Figure 2.3 For any image, the values of higher bit-planes define the perceptual significance of the image, whereas the lower bit-planes describe the smoothing parts. In Figure 2.4, we can see how a grayscale image is portrayed individually by the bit-planes. Here, it is also found that the higher bit-planes are organized in a unique manner, providing a proper shape to the image. But, in the lower bit-planes, bits are appear at random. This is the reason for preferring least significant bit (LSB) planes in imperceptible data hiding practice. This issue is comprehensively discussed in Section 2.2.1.

It is reasonable to ask why all images are considered to be grayscale instead of color images, whereas, in multimedia data transmission, visual objects are mostly in the form of color. Also, the human visual system generates a color impression of any perceived entity. Nearly 6.5 million cone cells in the human eye are responsible for color vision. Approximately 62%–65% of these cones are predominantly sensitive to red light, 30%–33% are responsive to green light, and only 2%–3% for blue, although the 'blue cones' are the most sensitive. Visible light occupies 0.40 µm to 0.70 µm of the electromagnetic spectrum. For light near 0.70 µm, the maximum absorption is performed by the 'red cones,' whereas

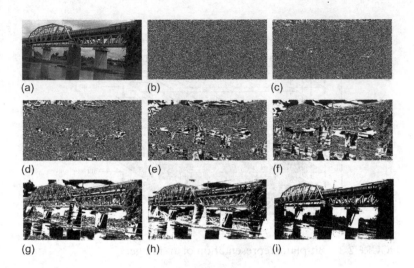

FIGURE 2.4 (a) Original image; (b) 1st bit (LSB) plane; (c) 2nd bit-plane; (d) 3rd bit-plane; (e) 4th bit-plane; (f) 5th bit-plane; (g) 6th bit-plane; (h) 7th bit-plane; (i) 8th (MSB) bit-plane.

a lower number are absorbed by 'green cones.' Similarly, near 0.40 μm maximum absorption is performed by blue cones and less by green cones. Light with a frequency near 0.54 μm is highly absorbed by green cones, but this absorption gradually decreases with both rising and falling frequencies. A similar absorption pattern is experienced by both the blue and red cones, where maximum absorption is performed at 0.45 μm and 0.58 μm for the blue and red cones, respectively. Because of this type of absorption feature, the human eye is able to perceive a great number of colors in the form of various combinations of red, green, and blue colors. A similar concept is followed in digital color image representation.

From the use of paints in childhood, we all know that red, yellow, and blue are the three primary colors, and all other colors can be produced from the proportionate mixing of these primary colors. Similarly, in digital color image depiction, any image pixel value is described by a unique combination of some basic color plane values. Based on the color-defining planes, different color models, such as RGB (Red-Green-Blue), CMY (Cyan-Magenta-Yellow), and HSI (Hue-Saturation-Intensity), are found in the digital image processing domain. In this book, the RGB model, being the most widely used, has been chosen to illustrate color image representation. Here, red (R), green (G), and blue (B) are the basic color planes. Each plane, normally defined to be a 256-level plane, is analogous to the grayscale image plane, i.e. every plane may produce 256 shades with values within the range of 0 to 255. Therefore, the three planes together can generate $(256)^3 = 16,777,216$ bit-depth combinations, each of which results in a distinct color shade available in the visual spectrum. This RGB model is elaborated in Figure 2.5 using Cartesian coordinates.

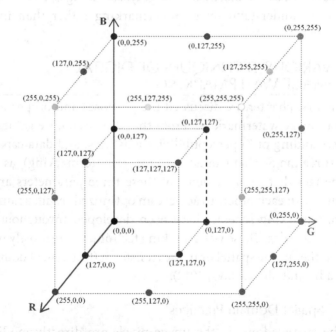

FIGURE 2.5 Cartesian representation of color pixels considering three primary colors– red, green and blue as three coordinates R, G and B respectively. (For original color image, see: https://docs.google.com/document/d/1YGd5aytCI8ThsZGeKSOTiTEJO5cFQKYF1a87kPHp9_0/edit?usp=sharing.)

The three coordinate axes are labeled red, green, and blue, and every point in the coordinate cube generates a distinct color, where the point location basically stands for the particular pixel bit-depth.

Figure 2.6 shows the individual color plane contributions and their corresponding grayscale representations for a color image using the RGB model. From this image, it is clear that any color image can be decomposed into different image planes and each plane is exactly analogous to a grayscale image. Considering the color planes individually, all the operations regarding image processing or watermarking in color images will be similar to those of the grayscale images. In digital watermarking, the blue plane is the one most usually modified. Therefore, for convenience, this book continues the discussion for grayscale images, as our main aim is to understand image watermarking rather than image processing.

2.2 VARIOUS TECHNIQUES OF DIGITAL IMAGE WATERMARKING

In the first chapter, we discussed the purposes and properties of digital image watermarking. From that discussion, we gained an understanding of imperceptibility, robustness, and data capacity (the three major characteristics of image watermarking), as well as the fact that improvements of these three parameters are in conflict with each other. To achieve an optimized result, a number of watermarking schemes have been developed throughout the last few decades. These watermarking techniques are mostly practiced either in the spatial or in the transform (frequency) domains (Kavadia and Shrivastava, 2012).

2.2.1 Spatial Domain Practices

In the spatial domain, the image pixels are directly modified by the watermark image pixels to produce the watermarked image. One of the most conventional spatial domain-based image watermarking techniques is the least significant bit (LSB) modification (Goyal and Kumar, 2014). This modification may

FIGURE 2.6 (a) Original image; (b) red-plane contribution; (c) green-plane contribution; (d) blue-plane contribution; (e)–(g) corresponding 256-level single plane representations for red, green and blue planes. (For original color image, see: https://docs.google.com/document/d/1HE9p BaHPkT54Cj7yEfhbP7jm2HLyRQzesKI9PXvqZRc/edit?usp=sharing.)

be performed simply by replacing the LSBs of cover image pixels by the watermark bits. Conditional bit replacements, like LSB matching (Xu et al., 2010) or adaptive bit replacement techniques (Sinha Roy et al., 2018), exemplified in the next section, are also included in this category. LSB modifications can also utilize a block-based approach by performing the watermarking operations after splitting the original image into several blocks or regions. As we have seen in Figure 2.4, in the LSB plane bits appear haphazardly. Due to this randomness, modification in this plane remains perceptually undetectable. Moreover, quantitative errors in pixel value decrease for LSB modification with respect to MSB (most significant bit) modification. The maximum difference obtained is 1 (2^0) for bit reversal in the LSB plane, whereas in the MSB plane the maximum difference is 128 (2^7). Thus, LSB coding is preferred when trying to achieve high imperceptibility, and MSB coding is preferred for visible watermarking. Later we find that, when using LSB modification, robustness is very poor. Moreover, a noise component is almost always found when the watermark is retrieved. To date, LSB is usually selected for its simple execution with high-quality data transparency. As an example, in 2013, a very simple digital image watermarking process using an LSB replacement procedure was described by G. Kaur and K. Kaur (2013), as it was very straightforward to implement and gave good imperceptibility, even though its robustness was not remarkable.

The patchwork technique is another type of spatial domain watermarking, where the cover image is divided into two subsets and opposing operations are applied to these. For example, if a factor is used to *increase* one of the subset values, the same factor will be used on the other subset to *decrease* the subset value. When retrieved, the difference between the two subsets determines the presence of the watermark. If the difference is twice the increasing factor, it is the watermarked image; otherwise, it is assumed that the watermark is not present in it. This technique could also be useful for the purpose of copyright protection.

Some correlation-based watermarking methods can also be performed in the spatial domain. A typical process is to convert the watermark into a pseudo-random noise and add it to the cover image. The validity of the watermark in an image is examined by finding the correlation between the random noise and the image at the receiving end. If the correlation result exceeds a certain threshold value, the watermark is detected, and otherwise, it is not. In predictive coding, another spatial watermarking practice, the correlation between adjacent pixels, is used: a set of pixels where the watermark is to be embedded is chosen, and alternate pixels are replaced by the difference between their adjacent pixels. This can be further improved by adding a constant to all the differences. Often a cipher key is provided for the receiver to extract the embedded watermark. These techniques offer much greater robustness when compared to that of simple LSB coding.

Very often these spatial domain watermarking methods make use of the imperfect nature of HVS. The word 'imperfect' is used here as HVS does not give equal importance to the whole visual stimuli and cannot perceive any minor changes in aesthetics. An HVS-dependent watermarking technique is a method to insert a large amount of information into those regions of an image that are beyond human perception. Although the pixel modifications are performed in the spatial domain, some frequency domain filtering processes are often employed in this type of technique in order to recognize the regions of interest (ROI) (Lee et al., 2005). In Chapter 3, HVS-based watermarking is discussed in detail together with the concepts of saliency detection and just noticeable distortion (JND) (Li et al., 2012b). These HVS-based methods are capable of providing optimized results with regard to robustness, data transparency, and payload capacity.

2.2.2 Frequency Domain Practices

Most signal processing actions are frequently performed and mathematically expressed in an exact field, known as a frequency domain. The frequency of an image signal refers to the rate of change in intensity values of the image pixels. The high-frequency components are consistent with the edges present in an image and the low-frequency components correspond to the smoothing prefectures, or regions, of that image, as shown in Figure 2.7. Here, the original image, given in Figure 2.7a, is reconstructed individually by the high-frequency components and by the low-frequency components in Figure 2.7b,c, respectively.

Frequency domain watermarking techniques generally insert the authentication mark into an image by modifying the higher-frequency components in order to enhance the robustness of the mark. Moreover, modifications in high-frequency regions are less noticeable as the HVS is less sensitive at the edges. A lot of researchers have been working to develop various watermarking algorithms based on different frequency domain approaches. This frequency domain work is principally categorized by its basic operation, known as image transformation, through which an image can be represented in a frequency domain. Thus, frequency domain techniques are often referred to as transform domain techniques. Among many others, three transformation techniques – discrete Fourier transform (DFT) (Li et al., 2012), discrete cosine transform (DCT) (Patra et al., 2010; Shaikh and Deshmukh, 2013), and discrete wavelet transform (DWT)

(a)   (b)   (c)

FIGURE 2.7   (a) Original image; (b) reconstructed image through high-frequency coefficients; (c) reconstructed image through low-frequency coefficients.

(Majumdar et al., 2011; Shah et al., 2015; Lin and Lin, 2009) – have been the main ones used.

Jean-Baptiste Joseph Fourier was the first mathematician to state that any periodic signal could be expressed by a series of sinusoidal waves. This concept led to the development of Fourier series to define a periodic signal $x(t)$, with time period T, by a linear combination of harmonically allied complex exponentials as,

$$x(t) = \sum_{k=-\infty}^{\infty} c_k e^{j\frac{2\pi k}{T}t} \qquad (2.1)$$

where,

$$c_k = \frac{1}{T} \int_{-T/2}^{T/2} x(t) e^{-j\frac{2\pi k}{T}t} dt, \quad k = 0, \pm 1, \pm 2, \pm 3 \ldots \qquad (2.2)$$

An aperiodic signal could be treated as a periodic signal by letting time T tend to infinity. Now, for any signal $x_C(t)$, with time period $T \to \infty$, Eq. (2.1) can be modified as shown below,

$$x_C(t) = \frac{1}{2\pi} \int_{-\infty}^{\infty} X_C(\Omega) e^{j\Omega t} d\Omega \qquad (2.3)$$

where, $\Omega = \dfrac{2\pi k}{T}$ and $X_C(\Omega)$ is a function of the continuous variable Ω. From Eq. (2.3) we obtain,

$$X_C(\Omega) = \int_{-\infty}^{\infty} x_C(t) e^{-j\Omega t} dt \qquad (2.4)$$

where Ω is inversely proportional to time T. Hence, the expression $X_C(\Omega)$ must lie within the frequency domain. The mathematical operation described in Eq. (2.4) represents any continuous time function $x_C(t)$ as a function of its frequency components and is known as continuous time Fourier transform (CFT). Eq. (2.3) defines the analogous inverse Fourier transform to reconstruct the time domain signal from its frequency components.

Present-day data transmission raises with the digital signals, which are defined in discrete instants of time. Discrete time Fourier transform (DTFT) gives the Fourier transform of a discrete time function $x_d(t)$ as a continuous function of Ω. But, in practical terms, we need to deal with a finite number of samples in the frequency domain. Taking N to be the number of samples of the DTFT output function over a single time period, we get the discrete Fourier transform (DFT) of the discrete time signal $x_d(t)$ as,

$$X_d(v) = \sum_{n=0}^{N-1} x_d(n) e^{-j\frac{2\pi v}{N}n}, \quad v = 0,\ 1,\ 2 \ldots N-1 \qquad (2.5)$$

where $X_d(v) = X_d(\Omega)\big|_{\Omega = \frac{2\pi v}{N}}$ and $x_d(n) = x_d(t)\big|_{t=nT_s}$ for sample time T_s. The inverse discrete Fourier transform (IDFT) is defined as,

$$x_d(n) = \frac{1}{N} \sum_{v=0}^{N-1} X_d(v) e^{j\frac{2\pi n}{N}v}, \quad n = 0,\ 1,\ 2 \ldots N-1 \qquad (2.6)$$

In Section 2.1, we saw that a digital image is considered as a two-dimensional array of distinct pixel values. Therefore, the Fourier analysis for an image is performed by employing a two-dimensional DFT. For a grayscale image f_i of size M × N, the DFT is obtained through Eq. (2.7). The spatial image comes back from its corresponding Fourier transform domain through a two-dimensional IDFT, given in Eq. (2.8).

$$F_i(u,v) = \sum_{m=0}^{M-1} \sum_{n=0}^{N-1} f_i(m,n) e^{-j2\pi\left(\frac{um}{M} + \frac{vn}{N}\right)}, \quad \begin{array}{l} u = 0,\ 1,\ 2 \ldots m-1 \\ v = 0,\ 1,\ 2 \ldots N-1 \end{array} \qquad (2.7)$$

$$f_i(m,n) = \frac{1}{MN} \sum_{u=0}^{M-1} \sum_{v=0}^{N-1} F_i(u,v) e^{j2\pi\left(\frac{um}{M}+\frac{vn}{N}\right)},$$

$$\tag{2.8}$$

$$m = 0, 1, 2 \ldots m-1$$
$$n = 0, 1, 2 \ldots N-1$$

In practice, a DFT operation is generally performed using an FFT (fast Fourier transform) algorithm. Readers should keep in mind that an FFT is not *any* transformation but a specific technique to obtain a DFT. For an M × N gray image (Figure 2.8a), its frequency domain representation is obtained through a DFT (using the command 'fft2' in MATLAB), shown in Figure 2.8b. This is

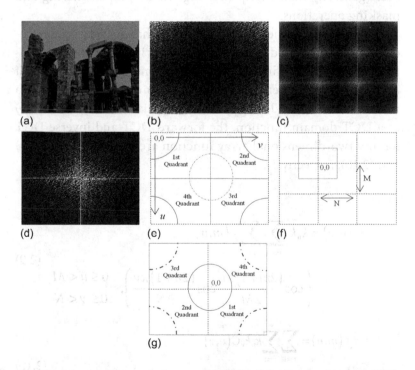

FIGURE 2.8 (a) Grayscale image; (b) DFT of the image; (c) periodicity of DFT; (d) DFT of the image taking account of the DC component at the center; (e)–(g) simplified illustrations for Figures (b)–(d) respectively.

illustrated in Figure 2.8e. $F(0,0)$ is basically the sum of all pixel values and is often denoted by the DC component of a Fourier transform because for $F(0,0)$, the exponential term becomes unity. As described in Figure 2.8e, the low-frequency components are produced at the corners of the transformed image, whereas the high-frequency components are at the central area. The periodicity of two-dimensional DFT states that the DFT of any image of size M × N repeats itself after every M (row-wise) and N (column-wise) points, which is exemplified in Figure 2.8c,f. For our own purposes, we can shift the origin at the (0,0) point to generate quadrants as shown in Figure 2.8d,g. Digital watermarking is performed in the DFT domain by embedding information into the best (generally high-) frequency regions by proper filtering and masking operations.

The discrete cosine transform (DCT) has been used for frequency domain-based signal processing, as it offers several advantages over DFT. The first and foremost advantage of DCT is that it eradicates the requirement to deal with complex numbers in data manipulation. Considering $f(m,n)$ as a spatial function and $C(u,v)$ as a DCT domain function, the forward DCT and inverse DCT for any two-dimensional array function are defined by Eqs. (2.9) and (2.10), respectively.

$$
C(u,v) = k_u k_v \sum_{m=1}^{\infty} \sum_{n=1}^{\infty} f(m,n)
$$
$$
\left(\cos \frac{(2m+1)\pi u}{2M} \cos \frac{(2n+1)\pi v}{2N} \right), \quad \begin{array}{l} 0 \le u < M \\ 0 \le v < N \end{array}
$$
(2.9)

$$
f(m,n) = \sum_{u=1}^{\infty} \sum_{v=1}^{\infty} k_u k_v C(u,v)
$$
$$
\left(\cos \frac{(2m+1)\pi u}{2M} \cos \frac{(2n+1)\pi v}{2N} \right), \quad \begin{array}{l} 0 \le m < M \\ 0 \le n < N \end{array}
$$
(2.10)

where,

$$k_u = \sqrt{\frac{1}{M}}, \quad u = 0 \qquad \qquad k_v = \sqrt{\frac{1}{M}}, \quad v = 0$$

$$\text{and}$$

$$= \sqrt{\frac{2}{N}}, \quad 1 \le u < M \qquad \qquad = \sqrt{\frac{2}{N}}, \quad 1 \le v < N$$

DCT is also significant for improved energy compaction, i.e. the capability to bundle the spatial energy into as few frequency coefficients as possible. This property facilitates image compression as well as embodying an image by only a few frequency coefficients instead of the entire set of pixels. This is because the maximum amount of the signal energy lies within the low-frequency regions and signal values representing high-frequency components are often too small to be wiped out with minute visual distortion. In general, two-dimensional DCT is achieved through 8 × 8 DCT by applying a fast 8-point DCT algorithm to each row followed by each column. The 8 × 8 discrete cosine basis functions, generated in MATLAB, are shown in Figure 2.9a. Similar to DFT, in the DCT domain, the origin, or the (0,0) point, is also at the top-left corner. But unlike DFT, the low-frequency components gather around the DC component and the image frequency increases diagonally. The DCT representation of the image, given in Figure 2.8a, is shown in Figure 2.9b. The high-frequency regions are mostly preferred for embedding a watermark.

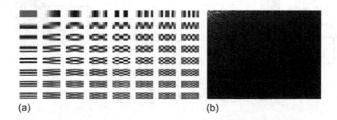

(a)   (b)

FIGURE 2.9   (a) 8 × 8 discrete cosine basis functions; (b) DCT for the image shown in Figure 2.8a.

Discrete wavelet transform (DWT) deals with both the frequency components and the temporal details. Here, basis functions vary both in frequency and spatial range. In Fourier analysis, only the frequency components are under consideration, which is another drawback for DFT with respect to DWT. According to Heisenberg's principle of uncertainty, high-frequency resolution cannot be achieved with improved temporal resolution. Therefore, the wavelet transform is designed in such a manner that good frequency resolution is obtained for the low-frequency components and the temporal resolution is superior for the high-frequency components.

In wavelet analysis, high-pass filters (HPF) and low-pass filters (LPF) are used to decompose the frequency components of a signal into four sub-bands, as shown in Figure 2.10. Here we can see that out of these four sub-bands, one (LL) is the approximation sub-signal, providing a maximum amount of signal energy as well as a general trend to pixel values. The remaining three, i.e. HL, LH, and HH, are horizontal, vertical and diagonal sub-bands holding some details like edges. A reverse path is followed to get the inverse DWT (IDWT).

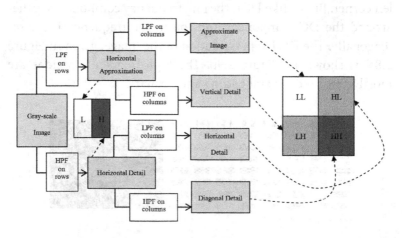

FIGURE 2.10 Generation of sub-signal bands for grayscale images through two-dimensional wavelet transforms.

L-level decomposition is achieved by applying this technique L times to successively generated approximation sub-signals and, as a result, 3L + 1 sub-bands are finally obtained. A multilevel wavelet transform is demonstrated in Figure 2.11a. For a grayscale image (Figure 2.11b), the approximation and detailed sub-images obtained through a 3-level wavelet transform are shown in Figure 2.11c. As the edges are less noticeable to HVS, a watermark can be embedded into the HH sub-bands after obtaining multilevel thresholds. Thus, watermarking in DWT (Totla and Bapat, 2013) offers improved robustness with good quality, visual transparency.

We have seen that the transform domain watermarking schemes can offer enhanced robustness with tolerable data transparency. Still, spatial watermarking is important as in the spatial domain a larger payload can be applied. Moreover, hardware

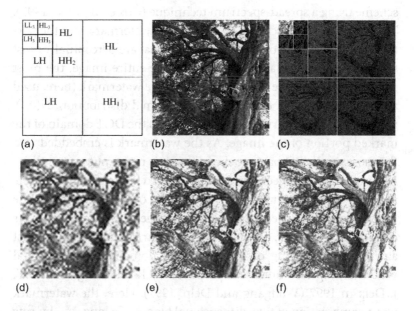

FIGURE 2.11   (a) Two-dimensional DWT multilevel decomposition; (b) grayscale image; (c) DWT 3-level decomposition of that shown in (b); (d)–(f) approximations (128 × 128, 256 × 256, 512 × 512) achieved from (c) at different levels.

implementation for spatial methodologies is less complex and the computational cost is also very low with respect to the frequency domain techniques. In the next section, various watermarking schemes developed in both domains are briefly described.

2.3 SOME EXISTING FRAMEWORKS FOR DIGITAL IMAGE WATERMARKING

The concept of electronic or digital watermarking was first introduced by Tanaka et al. in 1990 (Tanaka et al., 1990) and it was initially brought into use by Tirkel et al. (1993). Since that time, a number of watermarking algorithms have been put in practice as a copyright protection tool. Some of these methods, which have been developed over the years either in the spatial or in the frequency domain, are given in this section.

J. Cox et al. introduced a frequency domain watermarking scheme using a spread-spectrum technique (Cox et al., 1996, 1997). The intention of the scheme was to insert a watermark into the particular spectral components of the signal that are perceptually most significant. By calculating the DCT of the entire image, the most significant regions are marked out. Then the watermark (here, used as a sequence of real numbers with a normal distribution, $N(0,1)$, with 0 mean and 1 variance) is inserted into the DCT domain of the marked portion of the image. As the watermark is embedded into the most significant parts of the image, it is resistant to general signal processing and geometric distortion. To make it imperceptible in any frequency beam, Cox et al. spread it over a broad band. The drawback of this scheme is that it needs the original image for its extraction. Moreover, the author did not clarify whether it is robust against photocopying or not. A spatial domain invisible digital watermarking technique was proposed by R. B. Wolfgang and E. J. Delp in 1997 (Wolfgang and Delp, 1997). Here, the watermark was a combination of two-dimensional blocks of a long row-by-row sequence having the same image size. In this scheme, the author additionally used a testing paradigm with different ranges through which any type of forgery affecting the image could be determined.

In 1998, a suitable spread-spectrum dependent watermarking technique was developed by Hung and Shi (1998). For this technique, watermark insertion was performed based on human perception, i.e. the strength of watermark was adopted by considering the human visual system for the texture and brightness sensitivity. To measure this, a JPEG quantization table was used. In the DCT domain, the DC components indicate the brightness sensitivity, and the texture sensitivity is measured by quantizing the DCT coefficients. Then all the image blocks are divided into three classes according to their brightness and texture values. These are (i) dark and weak texture, (ii) bright and strong texture, and (iii) the remaining portion. Finally, three watermark sequences are embedded in the lower-frequency coefficients of each block.

Another HVS-based watermarking technology was introduced by Kim, Kwon, and Park in 1999 (Kim et al., 1999). But here, instead of DCT, wavelet transform was used. In this proposal, the energy of each wavelet band is calculated and, based on these results, the watermark sequence changes proportionally. The changing rate of the sinusoidal pattern per subtended visual angle (unit cycles per degree) gives the estimate of the image characteristics as well as the visual weight of the watermark in each wavelet transform band. In the same year, M. Kankanahalli et al. (1999) proposed a visible watermarking technique based on the DCT domain. Unlike Cox et al., instead of calculating the DCT of the whole image, Kankanahalli et al. initially clustered the image into several blocks and then evaluated the DCT for each block. Then the authors classified the blocks into six types according to noise sensitivity. Finally, the DCT coefficients of the watermark are embedded into the DCT coefficients of each block of the host image and hence produce the watermarked image. A transform domain technique was chosen again by W. Zhu et al. (1999). Instead of DCT, DWT is employed here. In this case, the watermarking algorithm was developed by inserting the watermark into all high-pass wavelet coefficients, and the watermark, which is basically a Gaussian function, can be detected at different resolutions.

Parallel to these transform domain frameworks, N. Nikolaidis and I. Pitas developed some spatial domain techniques to embed a little more information through watermarking (Nikolaidis and Pitas, 1996, 1998). I. Pitas (1996) proposed an image watermarking method where the cover image pixels are divided into different regions or sets. The watermark, considered as a combination of an equal number of 1s and 0s, is embedded into one of the sets. Therefore, the intensity level of the pixels in one set remains the same whereas, in the other one, it is altered so that the watermark can be detected at the extraction end by comparing the intensity. The main advantage is that this is a blind watermarking scheme, i.e. the original image is not required in watermark detection process. Moreover, it ensures better resistance to JPEG compression and low-pass filtering.

A fragile digital watermarking scheme was also proposed in 1999 by D. Kundur and D. Hatzinakons (1999). Their purpose was to detect tampering in digital images. It was a wavelet transform-based approach, where the watermark is inscribed in the host image through a quantization of corresponding coefficients. As a discrete wavelet is introduced, it provides both the spatial localization and a frequency spectrum of the watermark in the cover image. Thus, the ability to identify the specific regions of the watermarked image that have passed through tampering is gained and a global spreading of the mark makes it impressible to large-scale signal distortion. The authors argued that this space-frequency distortion-based modification is more effective and practical for tamper-proofing and also offers good credibility.

In the spatial domain, S.P. Mohanty et al. worked with both visible and invisible watermarking to be used on the same image, calling their approach 'dual watermarking' (Mohanty et al., 1999). The invisible watermark acts as a backup for the visible watermark. In this scheme, before embedding, authors clustered both the host image and the watermark image in blocks of the same size, although it is not mandatory for both host and watermark images to be of the same size. The mean and variance of the image blocks are then calculated. According to these results, scaling and embedding factors are added. The range of the maximum and minimum values of the mean and median are chosen as the parameters to use. The watermark is considered as a pseudorandom binary sequence arranged into blocks of size 4 × 4 or 8 × 8. The watermark bits are XOR-ed with a particular bit-plane of the original image and the modified bit-plane is merged with the other bit-planes to generate the watermarked image. The bit-plane is chosen such that the signal-to-noise ratio (SNR) does not exceed an endurable threshold value. Hence imperceptibility is optimized in this dual watermarking process.

In 2000, Chen et al. proposed another watermarking scheme (Chen et al., 2000), embedding a binary image as a watermark through the DCT approach. The watermarked image is imperceptible to the human visual system. During the embedding and extracting process a feature-based method is used to locate the watermark. The quantity of watermark bits is of the key consideration in this procedure. A spatial domain-based image watermarking system was put forward by A. Nikolaidis and I. Pitas (2001). They chose a chaotic watermark and developed their watermark embedding method based on sectoring the host image and locating the segments that are robust to various image manipulations. It has been shown that this algorithm offers improved robustness to several types of attacks, such as filtering, scaling, compression, rotation, and cropping. H. Wood (2007) worked with a grayscale watermark to be inserted into a color image through two distinct domains – the spatial domain and the discrete cosine transform domain. A comparative study between these two schemes shows that both domains offer invisibility of the watermark, when robustness is considered, the watermark was more robust in the DCT domain than the spatial domain. Thus, this work allowed the conclusion that a DCT domain offers higher robustness against signal processing attacks than a spatial domain. C.

Tsai et al. (2005) set up a reversible practice of information hiding for binary images where a lossless reconstruction of the image is made using pair-wise logical computation.

For the authentication of crime scene images, A. T. S. Ho et al. introduced a semi-fragile watermarking system (Ho et al., 2006) using pinned sine transform (PST). By this method, the maliciously tampered sections of image are detected with high validity. This watermarking technique is sensitive to any kind of texture variation in the watermarked images, and this property is of great importance for crime scene image authentication. Another consideration is a fragile watermarking scheme, which can detect any type of malicious modification performed to a database. An innovative fragile watermarking scheme was designed by H. Guo et al. in 2006 (Guo et al., 2006). In this scheme, the modifications applied to the database are able not only to be detected by the embedded watermark but also to be localized and even characterized. In this same year, another fragile watermarking scheme for a three-dimensional model was introduced by C. M. Chou and D. C Tseng (2006). This data authentication proposal relied on the sensitivity of vertex geometry. In this method, neither the original model nor the watermark is required for authentication.

Some ROI-dependent watermarking methods were proposed by Ni and Ruan (2006) and Fan et al. (2008), where watermarks were embedded into interested regions of the cover object. ROI recognition was performed in these algorithms through user-defined functions. Instead of selecting the regions of interest by user, Mohanty and Bhargava (2008) made use of various features of HVS, such as intensity, contrast, sharpness, texture, and location in their watermarking scheme.

In 2008, H. Wang et al. (2008) proposed another chaotic watermarking scheme for authentication of JPEG images. The quantized DCT coefficients after entropy decoding are mapped to the initial values of the chaotic system and then the watermark information, generated through this chaotic iteration, is embedded into the JPEG-compressed domain. Requantization operation does not invalidate tamper detection due to direct modification of the DCT coefficient after quantization. Extraction is performed by utilizing the same compression technique. Watermark extraction time and complexity of this method were claimed to be low.

In 2009, G. Roslin Nesa Kumari et al. (2009) proposed a new LSB modification-based digital image watermarking technique. For this technique, the watermark is embedded in areas of the same size, depending upon their grayscale values and coordinate positions. Also, in 2009, Sur et al. utilized HVS to improve data transparency through an adaptive watermark embedding process in the spatial domain (Sur et al., 2009). They inserted the watermark into the least salient regions of the cover image so that the modifications would not be perceived in any aesthetic sense. The data capacity of this technique was low with respect to other spatial domain techniques. Chen et al. proposed another spatial domain fragile watermarking technique (Chen and Wang, 2009) that utilizes a Fuzzy C-means (FCM) clustering. In addition, the embedding procedure is practiced in such a way that this technique is also good for tamper detection. It is based on the concept of incorporate block-wise dependency of information in an embedding process to prevent vector quantization (VQ) attack without compromising the localization capabilities of the scheme. In the following year, a vector quantization-based watermarking technique was developed by J. Anitha et al. (2010), where a self-organizing feature map (SOFM) is used. The proposed system uses the codebook partition technique to embed the watermark into selected VQ-encoded blocks. Thus, the watermark is present in both the VQ-compressed image and the reconstructed image, forming a significant feature of this process. As a result of this type of watermark embedding technique, the original is not required in the watermark extraction. The experimental results, shown in their paper, reveal that it offers good robustness against several signal processing attacks. An innovative spatial domain digital watermarking technique that can be used for mages with few colors, called Spatial Unified Key Insertion (SUKI), was proposed by

Shu-Kei Yip et al. (2006). The proposed algorithm has a high pay-load and excellent resistance to JPEG attack. It has a better BER (bit error rate) (as low as zero) and PSNR (peak signal-to-noise ratio). Moreover, no additional color is introduced and the palette remains unchanged after watermark embedding.

In 2011, Mohammed and Sidqi introduced an image water-marking scheme based on the multiband wavelet transformation method (Mohammed and Sidqi, 2011). Initially, the proposed scheme is tested on the spatial domain in order to compare its results with the frequency domain. In the frequency domain, an adaptive scheme is designed and implemented based on the selec-tion criteria of bands to embed the watermark. The disadvantage of the scheme is the involvement of a large number of wavelet bands in the embedding process.

Also, in 2011, a new image watermarking method was put forward by Bhattacharya et al. (2011). It makes use of both frag-ile and robust watermarking techniques. The embedded fragile watermark is used to assess the degradation undergone by the transmitted images. Robust image features are used to construct a reference watermark from the received image to evaluate the amount of degradation of the fragile watermark.

In the following year, 2012, F. Husain et al. introduced a digital image watermarking scheme (Husain et al., 2012) that combines the discrete wavelet transform and discrete fractional Fourier transform (DFRFT). This technique is treated as a non-blind method. The multi-resolution sub-band decomposition property of DWT is utilized and DFRFT is applied on selected sub-bands in the transform domain image. Compared with simple DWT- or DCT-based methods, this proposed system offers better robust-ness against most attacks, except for median filtering, AWGN (additive white Gaussian noise), and JPEG compression attack. In 2012, Kannammal et al. developed a medical image watermarking framework in which the electrocardiograph (ECG) and patients' demographic text ID act as double watermarks (Kannammal et al., 2012). By this method, the medical information of the patient is protected and any mismatching of diagnostic information is prevented.

A recent technique for watermarking has been developed using the salient properties of the image pixels. In 2011, A. Basu, T.S. Das, and S.K. Sarkar proposed an adaptive spatial domain image watermarking technique (Basu et al., 2011a) based on bottom-up Graph-Based Visual Saliency (GBVS). The embedding system described in this paper embeds the watermark into the uneven bit-depth salient image pixels. It provides better perceptual transparency to the human visual system. Moreover, it has an improved robustness when compared with the other existing spatial domain watermark embedding schemes. Another saliency-based watermarking technique described in the same year by Yaqing Niu et al. (2011) was proposed to assist image watermarking by producing a visual saliency modulated JND (just noticeable distortion) profile that can be used as a guide to optimize image watermarking. In this technique, JND is also considered in watermark embedding time rather than concentrate only in HVS. A further improvisation was made by A. Basu et al. (2011b), who developed a robust adaptive LSB replacement-based visual digital watermarking framework, which also employed HVS, to embed covert information into a gray image without any immediate visual distortion. Here, the cover image is divided into some sub-blocks, in which the watermark bits are inserted through an adaptive LSB replacement technique. HVS masking is utilized here by use of the luminance feature, entropy texture, and variance edge of the cover image. This approach offers an improved visual quality, higher payload, and resistance against several types of signal processing attacks.

In 2012, Yang Qianli and Cai Yanhong implemented a digital watermarking technique (Qianli and Yanhong, 2012) that utilized the two-dimensional discrete wavelet transform and cosine transform in watermark embedding. The authors worked with a gray image as the cover image, which had been transformed into the discrete wavelet domain three times and was intersected into sub-blocks. Every sub-block was transformed into the discrete cosine domain and was embedded with the watermarking elements, which were also transformed into the discrete cosine domain. The inverse transforms of both the domains were then performed over the watermarked image before transmitting. Results have shown that the technique ensures robustness against several types of signal processing attacks, such as JPEG compressing, low-pass filtering, and other noises. Later, the authors carried out the technique for color images by inserting the watermark into the red component only. It provides higher security as, after decomposing a received watermarked image, the watermark can be retrieved from the red component only (out of the three RGB components). Another new watermarking technique, this time proposed by B. Ram (2013), also deployed both the DWT and DCT in embedding the watermark. A pseudo-random sequence of real numbers is inserted into a selected set of DCT coefficients. Data implantation is performed in the frequency domain and the watermark is added to the selected coefficients ensuring significant image energy in the DWT domain in order to provide a high level of robustness. The exploratory results show that this method does not require the original image or the watermark in watermark extraction and that it is possible to pick out a good quality watermark from a watermarked image even after some key image processing attacks, such as JPEG compression, median filtering, average filtering, and cropping.

S. Veeramani and Y. Rakesh (2013) proposed an optimization-based image watermarking model, where the watermark is inserted into the host image in the Scalar Costa scheme (SCS) transform domain. The RC4 encryption algorithm is used here to generate a cipher form of the original image and the watermark is inserted into the less significant bit-planes of the cipher. This scheme ensures the ability to detect malicious tampering and also carries robustness against content-preserving processes such as compression, filtering, and salt-and-pepper noise. A biogeography-based optimization-dependent watermarking was performed in the DWT domain by R. Tiwari et al. in 2013 (Tiwari et al., 2013). This type of optimization is usually used to optimize the strength of a watermark for the purpose of spreading the watermark. Watermark pixels with different strengths are spread into different regions of the host image and are adaptively selected on the basis of having the least effect on the imperceptibility of that particular image. At the same time, C. Woo et al. developed a fragile watermarking technique for tamper detection (Woo and Lee, 2013). To identify the tamper location efficiently, a block-wise technique is used. By using a hash code for the blocks (final level) where the watermark is to be embedded and the for the blocks (upper level) where these are to be included in an image division process, a digital signature is constructed. This signature is treated as a watermark that has been randomly inserted into the selected image blocks. As a result, the tamper detection operation can be performed without thoroughly testing the watermarked image blocks.

An adaptive digital image watermarking method, utilizing fuzzy logic and Tabu Search, was proposed by A. Latif in 2013 (Latif, 2013). In this framework the host image is divided into different blocks and, after applying parametric Slant-Hadamard transforms to each block, the watermark is embedded into each block in that transform domain. The proposed transform is useful to enhance robustness, and utilization of the fuzzy gradient model ensures imperceptibility. Thus, this method overcomes the conflict between robustness and imperceptibility. A wavelet-based watermarking scheme was introduced by H. B. Razafindradina and A. M. Karim (2013). In this proposed algorithm the watermark is planted into the host image by adding edge in the HH subband of the host image after wavelet decomposition. This method is described as a blind and flexible algorithm. The flexibility exists because of some consistent parameters, including modification effects on the improvement of robustness and hiding capacity.

To improve the watermark payload, S. Banerjee proposed a new color image watermarking algorithm (Banerjee et al., 2015) based on the residue number system (RNS). RNS, developed from the Chinese remainder theorem of modular arithmetic, refers to a large integer dependent on a set of smaller numbers. In this watermarking procedure, the pixel values from three different watermark images are taken and inserted into the cover image. From the empirical results, the author concluded that the watermarks were imperceptibly embedded into the cover image and successfully extracted from it, offering both better payload values and robustness.

Operating on the concept of region of uninterested (ROU), a feature of the human visual system, A. Basu et al. proposed a robust spatial domain-based adaptive image watermarking technique in 2013 (Basu and Sarkar, 2013). Here the watermark is embedded into the less significant pixels with respect to the visual attention model proposed by Itti and Koch (2000). It was shown that the perceptual transparency error after embedding the watermark is less than when using the normal LSB replacement technique.

The concept of a reversible data hiding scheme was revisited by Gui et al. in 2014 (Gui et al., 2014). An adaptive data insertion, rooted in generalized prediction-error expansion, is performed and an improved data hiding capacity is achieved through this process. Another reversible logic-based fragile watermarking algorithm was proposed by Chan et al. for hologram authentication (Chan et al., 2015). In this process, a watermark is masked into a hologram image in the transform domain. The resolution of the image is limited so that it offers visual transparency for the modified image.

A recent trend of image watermarking is to use singular value decomposition (SVD) and discrete wavelet transform simultaneously. In 2015, P. Shah et al. introduced this type of SVD- and DWT-based approach (Shah et al., 2015), offering improved robustness. In this technique, both the host image and the watermark are decomposed by using discrete wavelet transform followed by singular value decomposition in the LL band. The watermark values are subsequently injected through a scaling operation by identifying and substituting the singular values in every sub-band. In the same year, another DWT-based digital image watermarking process was put forward by Al-Nabhani et al. (2015). Here, a probabilistic neural network and a Haar filter are utilized, along with DWT, to embed a binary watermark in the preferred coefficient regions of a cover image. This approach considers robustness to be the main priority. When extracted the probabilistic neural network is again used to retrieve the watermark. From the experimental results, this method is successful in addressing the conflict between imperceptibility and robustness. Xiang-yang et al. proposed a robust digital watermarking scheme based on local polar harmonic transform (Xiang-yang et al., 2015). The watermark embedded through this algorithm can uphold itself against several signal processing noises as well as geometric deformations. In addition, the watermark does not affect the aesthetics of the original image after being inserted into it. A bimodal biometric validation-based technique was discussed by Wójtowicz and Ogiela (2016). In this method, both fingerprint and iris biometrics are treated as a watermark and embedded into the most suitable area of the cover image to produce the watermarked image. The testing outcomes confirm that the proposed system facilitates data validation with a significant precision level.

A blind digital image watermarking scheme was developed using a mixed modulation technique by Hu Hwai-Tsu and Hsu Ling-Yuan (2016). Here quantization index modulation and relative modulation are used in embedding a watermark. Principally, this is a discrete cosine transform domain-based approach. For the transform coefficients with very small estimated variation, the relative modulation is activated to implant the watermark. If the estimated deviation goes beyond a predetermined boundary threshold, the quantization index modulation is used to embed the watermark. Zhou Wujie et al. (2016) designed another fragile watermarking scheme for stereoscopic images. Here data embedding is processed using the perception of just noticeable differences and authentication is signified through tamper detection. The output results confirm that this method can provide enhanced security with a justified payload capacity and data transparency. A novel watermark decoding methodology was launched by in Sadreazami et al. in the contourlet domain (Sadreazami et al., 2016). For this technique, a multiplicative decoder was designed that utilized the standard inverse Gaussian probability density function as a prior requirement for the contourlet coefficients of the watermarked images.

Recently, in 2017, Susanto et al. (2017) described how DCT and HWT (Haar wavelet transform) can be utilized through a hybrid watermarking tool to address the issues regarding the divergence between imperceptibility and robustness. In this copyright protection framework, the original image initially undergoes HWT, and this step is followed by dividing selected LL sub-bands into 8 × 8 matrices. The DC coefficients of each of these matrices consist of information providing copyright protection. This approach provides robustness to some of the key signal processing attacks, such as JPEG compression or various filtering attacks.

As already discussed, the spatial domain watermarking schemes utilize the weaknesses of HVS to enhance imperceptibility. Saliency map-based techniques are categorized in this type, some of which have been discussed earlier. These watermarking tactics initially generate a saliency map of the original image and insert the maximum amount of information into the less salient regions. The bit insertion is generally performed using adaptive LSB replacement. A recent work in 2016, Basu et al. (2016), introduced a saliency-based watermarking algorithm, entirely developed in the spatial domain, by implanting the maximum quantity of data into the most salient regions. The experimental results reveal that data transparency for this technique is higher than the previous methods, but that the payload is less. In these saliency map-based watermarking models, developers generally compute a hiding capacity map (HCM) from the saliency map values to indicate the maximum payload for each cover image pixel. The generation of HCM can be performed automatically through intelligent techniques to achieve an optimized output. The subsequent chapters of this book are involved in executing an intelligent image watermarking technique where HVS characteristics are also made use of through saliency detection.

CHAPTER 3

Intelligent Systems Used in Copyright Protection

IN THE PRECEDING CHAPTERS, we have discussed several types of digital image watermarking and looked at their properties. As a copyright protection tool, the key advantage of digital watermarking is that the cover and the inserted watermark are largely inseparable so that the watermark is sustained within the cover, even after signal processing attacks or, indeed, even after the conversion of the cover into a different file format. Moreover, the watermark can be made transparent, so that no degradation of the cover image will be detected in an aesthetic sense. From our previous discussion, we have learned that the capability of the embedded information or watermark to remain unaffected in an object is known as robustness; and this property is inversely proportional to imperceptibility, or the visual transparency of the watermark. Also, if the amount of data inserted is increased, it will be harder to make the watermark robust as well as imperceptible. Thus, these three features – imperceptibility, robustness, and

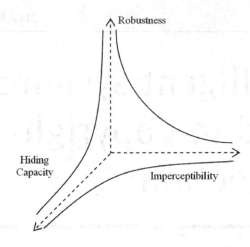

FIGURE 3.1   Relation between robustness, imperceptibility, and data hiding capacity.

data hiding capacity – are in conflict with each other (Figure 3.1). However, at the same time, these three attributes are all equally essential for any good copyright protection tool. It has already been found that spatial domain techniques can offer high-quality data transparency but are unable to provide a good level of robustness. In the frequency domain, robustness has been improved but payload capacity has decreased. This book introduces some intelligent techniques, such as saliency detection and image clustering, to embed a larger amount of information in such a manner that the embedded data remains imperceptible to the human visual system (HVS) and robustness is optimized.

Chapter 4 demonstrates how these intelligent techniques are utilized in a digital image watermarking scheme to provide an ultimate solution to resolve the trade-off between robustness, imperceptibility, and data capacity. This result is then confirmed through the system evaluation in Chapter 5. In this present chapter, we expand our discussion step-by-step to learn about these intelligent techniques and their usefulness in digital watermarking.

3.1 VISUAL SALIENCY AND ITS PURPOSE IN DIGITAL WATERMARKING

3.1.1 Saliency and Salient Object

First ,we need to know what is meant by 'saliency.' Saliency is the characteristic of a particle or living being through which it can stand out as being particularly noticeable or prominent over the neighboring objects. In other words, saliency is a feature of an object of being attractive with respect to its surroundings; and so that object is called a salient object. Similarly, the pixels or regions in an image will be considered to be salient if they are somehow special in terms of their shape, size, color, or in any other significant features when compared with their contiguous pixels (Toet, 2011; Carmi and Itti, 2006). Put simply, the regions in any visual stimuli that are most attractive to the HVS are treated as the most salient regions. As shown in Figure 3.2a, among all the balls, the black ones are the foremost perceived items and, therefore, the black balls are the most salient objects in this image. We can see in Figure 3.2b that the white balls can also be salient if placed on a black background. Hence, salient pixels in an image are those that are not regular in the sense of visual observation.

3.1.2 The Purpose of Saliency in Digital Watermarking

A perfect visual system is defined as one that has the capability to distinguish even the slightest changes in the visual stimuli. Invisible digital watermarking is possible only because of the

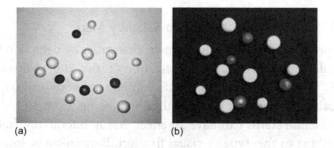

(a)   (b)

FIGURE 3.2   Salient objects.

restrictions of the human visual system, which is flawed according to this definition (Field and Brady, 1997). Moreover, the human visual system does not pay attention to all the visual stimuli; rather it is attracted first by the most salient regions, followed by the objects with less saliency. Thus, with respect to the saliency levels, image pixels can be divided into two major parts – a region of interest (ROI) and a region of noninterest (RNI). The insertion of a watermark into a cover image causes some modifications directly to the image pixels or its frequency components, and these modifications introduce some visual distortion to the cover object. But, if we can recognize the RNI pixels and embed the maximum information through modification of those particular pixels, the probability of any aesthetic degradation to be perceived by the HVS will be reduced. As we have already learned, in the frequency domain data transparency is improved by inserting the watermark through the alteration of a higher-frequency component, as the HVS is less sensitive to the edges. Now, the use of saliency detection benefits digital watermarking in two ways. First of all, a greater area is used to hide an increased quantity of data, hence providing high imperceptibility. Second, the ROI and RNI being indicated in the spatial plane, the insertion of a watermark can be performed in the spatial domain and, thus, it facilitates computational cost and time. In brief, we can say that the purpose of saliency in digital image watermarking is to enrich visual transparency with an increased payload, taking the full advantage of the imperfect nature of the HVS.

3.1.3 Saliency Detection and a Saliency Map

The previous discussion has shown that saliency detection basically depends on human visual fixation (Tatler, 2007; Chua et al., 2005), which is controlled by the complex nervous and psychological system of each individual human being. Even an individual's mental status can have an effect. Many interacting features contribute to this type of visual fixation. Bottom-up or low-level and top-down or high-level approaches are two broadly used feature-based algorithms (Borji, 2012) to sense the saliency in a visual stimulus using a computerized calculation. A simple example could be effective to distinguish between bottom-up and top-down behaviors. Such an example would be (Zhang et al., 2008; Koch and Ullman, 1985; Treisman and Gelade, 1980) analogous to the response of a child being attracted to a single red ball that has been placed together with some white balls. This prediction is obvious and no detailed mental state or behavioral account of that particular child is required. On the other hand, top-down attention is exemplified by the attraction of that child to a ball of their favorite color when a number of balls of several different colors are placed together. Here, the internal state, objective conditions, personal history, and personal experiences, would all need to be considered in order to formulate a prediction. Though the human visual system has an extraordinarily fast and reliable ability to detect a salient body, it is quite difficult to indicate the visually salient region through mechanical calculations. In the case of digital images, saliency detection through a bottom-up approach depends on the instantaneous sensory input, considering certain standard features, such as color, intensity, orientation, etc. In contrast, a top-down approach (Judd et al., 2009; Subramoniam et al., 2010; Kootsttra et al., 2008; Felzenszwalb et al., 2010) recommends several high-level features to direct visual attention. These high-level features include the face and facial components, human or other living things, cars, or text; and these types of prior knowledge are usually discovered throughout the life span of a human being. Certainly, the selection of these high-level features is very difficult and cannot be performed in a generic way. These top-down approaches can often lead to errors in saliency detection. For instance, a face detector-based algorithm can produce many false alarms for an image that does not contain any faces. Thus, for a standard and reliable process, digital watermarking mostly deals with bottom-up approaches. The next section provides a brief overview of some well-known bottom-up algorithms for saliency detection.

Note that the saliency map of an image is referred to its analogous image in which each pixel indicates the grade of saliency (generally within the range of 0 to 1) for relevant pixels in the original image. Figure 3.3 shows the saliency maps obtained through different saliency map-generating mechanisms for the image given in Figure 3.2a. Here, the brighter regions, having greater pixel values, stand for higher saliency and the dark pixels denote regions having less saliency.

3.1.4 Various Saliency Map Algorithms for Images

Visual perception through saliency detection is an emerging study involving several fields, such as computer vision, neuroscience, psychology, etc. Numerous computational models have been developed to generate saliency maps. A few of these, which are generic in nature and can be utilized in digital watermarking in a fast and effective way, are mentioned in this section.

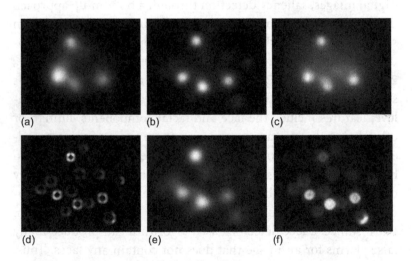

(a)   (b)   (c)

(d)   (e)   (f)

FIGURE 3.3 Saliency maps for the image shown in Figure 3.2a, obtained through (a) Itti's model, (b) the Simpsal method, (c) the GBVS approach, (d) the spectral residual method, (e) signature saliency, and (f) context-aware-based saliency.

Itti et al. introduced a visual attention model (VAM) (Itti et al., 1998) to spot the salient regions. In the VAM approach, intensity, color, and orientation disparity are mutually implicated to discover the regions of interest. This algorithm begins with a set of nine spatial scales, constructed using dyadic Gaussian pyramids (Greenspan et al., 1994), to generate horizontal and vertical image reduction factors by gradually applying low-pass filtering and sub-sampling to the input image. A group of linear center-surround operations, akin to the HVS receptive course, is deployed to differentiate between fine and coarse scales. This computation leads to the development of a distinct feature map for each feature. Finally, the overall saliency map is evaluated through the combination of normalized values of all the feature maps. Saliency detection through VAM is analogous to the normal visual phenomenon of HVS and is capable of sensing salient objects in visual stimuli by identifying spatial discontinuities. Later, the Simpsal method was developed and was brought into use in order to provide a simplified version of Itti et al.'s saliency detection algorithm.

J. Harel et al. developed a graph-based visual saliency (GBVS) model (Harel et al., 2006), which is also a bottom-up approach. Hence, only the instantaneous sensory input is under consideration in the saliency map computation (and not the inner state of the individual). This algorithm can be divided into two parts. In the first step, activation maps are developed on certain individual features. In the second step, a normalization of these maps is carried out in such a manner that the conspicuous regions are emphasized. Saliency prediction using GBVS aims to accentuate, in a visual stimulus, the regions that could be informative according to human visual fixation.

Information theory states that intelligent coding can effectively decompose an image signal into two parts, for instance, a novelty (innovation) part and a redundant or prior knowledge part. Hou and Zang (2007) proposed that the saliency map could be estimated through reassembling the innovation or novelty parts of an image by eliminating the statistically redundant components. They developed a spectral residual approach to distinguish the novelty and prior parts. This is basically a frequency domain-based practice that employs the two-dimensional fast Fourier transform (FFT) algorithm. Here, first, the log spectrum for the transformed image is computed and, then, the average shape of the log spectrum is approximated. Saliency is treated as the singularities obtained in a log spectrum with respect to the average or smoothed curve. This is a generalized model, i.e. it is for any kind of image, and it does not depend on prior knowledge regarding the image.

A figure–ground separation can be executed in the transform domain to trace the salient objects in an image. This methodology, called image signature, was put forward by X. Hou et al. (2012). It considers any grayscale image as a combination of foreground and background signals, which are sparsely supported, and occur respectively in the spatial domain and in the discrete cosine transform (DCT) domain. That means only a low number of non-zero components can be found in both of the signals, and so it is reasonably complicated to precisely segregate the foreground and background components for any image signal. This figure–ground approach considers the non-zero foreground components as regions of interest, and these can be approximately isolated by taking the sign of the transformed image components in the DCT domain. After computing the inverse DCT, the overall saliency map is obtained by smoothing the new squared image using a Gaussian kernel.

Additionally, S. Goferman et al. (2012) put forward the idea of context-aware-based saliency recognition. Instead of concentrating on the factors responsible for visual fixation on dominant objects, this approach gives greater attention to psychological factors. The pixels or regions of an image, through which the image may be defined psychologically, are considered as regions of interest.

Many more saliency detection algorithms have been developed during the last few decades, and improvisations are continuing to the present day. In digital image watermarking, we should try to employ a saliency map-generating algorithm that can form the saliency map in a fast and effective way, with the least reference to any statistical knowledge of input stimuli. This book deals with the spectral residual approach for generating a saliency map. This is a generalized model, i.e. it does not depend on the prior knowledge of the image. The spectral residual approach computes the saliency through a simple algorithm with an improved runtime. Here, saliency detection is executed by recognizing the busy regions; and later we will see that this property increases the hiding capacity as well as the data transparency. Moreover, when compared with the other models, saliency detection through the spectral residual approach is highly responsive to psychological patterns as well. The assimilation is shown in Figure 3.4. The first row of this figure consists of some salient objects in different forms, such as closure, curve, density, intersection, and inverse intersection, developed in five separate images. Several saliency detection methods are then applied to each of the images. The outcomes illustrate that salient object detection through the spectral residual method is much greater than through the others. In the next chapter, saliency map formation using the spectral residual-based method is discussed in detail.

3.2 INTELLIGENT IMAGE CLUSTERING: THE K-MEANS ALGORITHM

3.2.1 Data Clustering

Data clustering may be defined as a statistical categorization process used to place individuals of a population into different groups by working out quantitative comparisons of manifold features (Merriam-Webster Online Dictionary, 2008). The basic aim of data clustering is to reduce the magnitude and complexity of dealing with a large data set by considering the comparable objects as distinct groups. At the same time, it is also useful in identifying the salient features of an object. Technically, clustering is described as a process to split N observations into K groups depending on

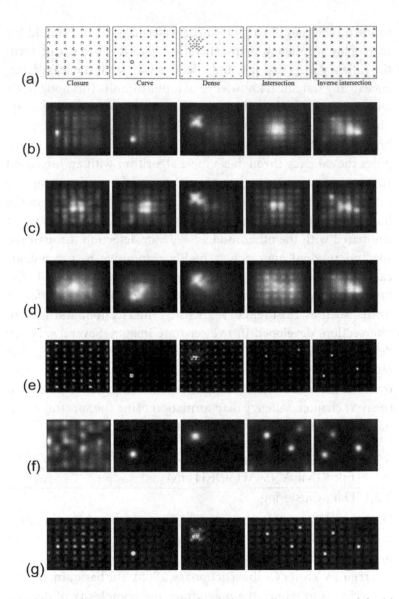

FIGURE 3.4 A comparative study for different saliency map models: (a) Different unknown patterns; (b)–(g) analogous saliency maps obtained through Itti's model, the Simpsal method, the GBVS model, the spectral residual model, signature saliency, and the context-aware-based saliency map model, respectively.

a few characteristic resemblances, such that the resemblances are high for objects in same group and low for objects belonging to different groups. But there are no such standard reference features for cluster analysis for any particular object type. As an example, if we intended to classify the animal kingdom, results would differ with respect to shape, size, nature, food habit, etc., i.e. different clusters are formed for different similarity features. Human beings are so intelligent that they can easily distinguish between observations in a rapid and multivariate way, whereas performing this as a computerized computation is a challenge.

Cluster analysis is effectively ubiquitous in biology, taxonomy, psychology, medical sciences, business and marketing, mathematics, engineering, computer vision, and many other research fields that need to deal with large data sets. For example, a well-known use of data clustering in the biological domain is genome data analysis (Baldi and Hatfield, 2002); or, in the interests of marketing efficiency, customers' needs and expectations could be addressed, after sorting them into several groups using suitable cluster analyses (Arabie and Hubert, 1994). Data clustering is also equally important for workforce management and planning (Hu et al., 2007), for example by grouping services and engagements. Both computer vision and artificial intelligence often exploit cluster analysis for image segmentation (Jain and Flynn, 1996; Frigui and Krishnapuram, 1999; Shi and Malik, 2000).

In the same way, clustering has become an essential tool for all types of multimedia data processing. Different types of clustering algorithms are mostly classified into two groups according to the property of nesting. These are partitional (or unnested) clustering and hierarchical (or nested) clustering (Jain, 2010). Partitional clustering minimally separates the input objects into a low number of non-overlapping subsets, so that each data type belongs to one and only one subset. In hierarchical-based clustering processes, a set of nested clusters are recursively formed, either in a divisive (top-down) or in an agglomerative (bottom-up) approach, and they are organized as a tree structure. In top-down models, clustering is initiated by considering all data points in the same cluster, and at each successive step, the splitting of clusters is performed until each cluster is optimized with a singleton data set. Agglomerative processes initially assume all data objects as individual clusters and the number of clusters is then reduced by merging the most comparable pairs of clusters. Single-link and complete-link, the two eminent examples of hierarchical clustering, are developed on the bases of minimum proximity (defining the shortest distance between two data points) and maximum proximity (defining maximum distance between two data points). The K-means algorithm is the most popular and broadly used partitional clustering model.

3.2.2 Need for Data Clustering in Digital Image Watermarking

Saliency mapping can guide an image watermarking scheme to insert the watermark in such an adaptive way that imperceptibility can be improved. The regions of interest and noninterest are distinguished through saliency detection and the efficiency of the watermarking algorithm is improved by increasing the payload in RNI portions and reducing the data hiding in ROI sections. But saliency mapping alone is not sufficient to perform this adaptive data insertion. This is because the saliency map pixel values for an image indicate how much saliency the corresponding image pixels have, and it is not possible to treat each and every pixel individually for amending its hiding capacity. For instance, a grayscale image of size 256 × 256 can contain 65,536 saliency map values, and thus it is difficult to prescribe such a high number of distinct hiding policies for a single image. This problem could be solved by applying some thresholds to the saliency map and dividing the image pixels into fewer clusters having distinct range of saliency. This would mean that the associated watermarking algorithm would only need to define a small number of hiding policies separately for each group, instead of treating all the image pixels independently. The grouping of saliency map pixels may be performed by defining the range of saliency manually. But, handled this way, the process will not be so generic. Because, for all types of images, saliency map pixel values will not be the same, in order to obtain a better data transparency and hiding capacity, it will be effective to vary the range according to the saliency map pixel values. A generic and optimum result can therefore be achieved by initiating an automated data clustering algorithm in preference to selecting the thresholds manually. This book now introduces the K-means algorithm as a data clustering mechanism that can be applied to saliency map pixels in order to enhance the data hiding efficiency of digital image watermarking schemes.

3.2.3 K-means Algorithm for Data Clustering

The K-means algorithm is one of the most conventional data clustering algorithms, developed more than sixty years ago. Its simple execution and high efficiency are what have made the K-means algorithm so well liked in various scientific fields over such a long period. The K-means algorithm basically clusters an input data set of N objects into K subsets, where the value of K is predetermined. During this partitioning process, this algorithm always tries to minimize the squared error between the experiential mean of a cluster and the data points in that cluster. In other word, the K-means algorithm aims to diminish the sum of the squared errors obtained for all the K clusters.

Let K be the number of clusters formed for a given data set with N the number of data points, then $P = \{p_n \mid n = 1, 2, \ldots, N\}$ and Q_1, $Q_2, \ldots, Q_K$ are the resulting subsets. If $\sigma_1, \sigma_2, \ldots, \sigma_K$ are the means for the clusters $Q_1, Q_2, \ldots, Q_K$, respectively, the sum of squared errors is defined as,

$$\Delta = \sum_{k=1}^{K} \sum_{p_n \in Q_k} \left\| p_n - \sigma_k \right\|^2 \tag{3.1}$$

Now, reducing the value of Δ is definitely a nondeterministic polynomial time (NP)-hard problem and that is why K-means can merely converge to a local optimum.

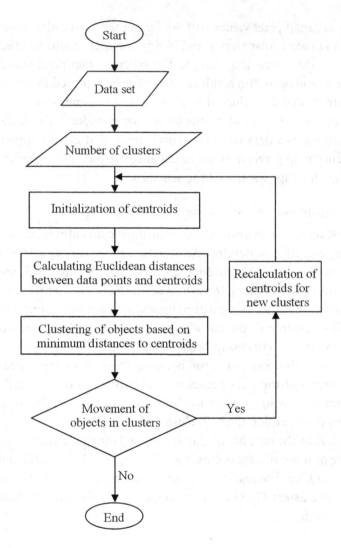

FIGURE 3.5 A flowchart for the K-means clustering algorithm.

The algorithm has two consecutive phases. In the first phase, the initial centroid or empirical mean σ_k, arbitrarily chosen from the data set P, is defined for each and every cluster Q_k, where, $k = 1, 2, \ldots, K$. The objective of the second phase is to determine the Euclidean distances between the centroids and all input data points and correlate these to the closest centroid. Thus, all the points are initially clustered into subsets through this premature grouping. This grouping is said to be premature because, in repeated steps, the means are reassessed to obtain an optimal result, and thus, centroid positions as well as cluster descriptions may vary. This successive evaluation eventually leads to stability for centroids; that means centroids remain almost same for any following estimates. This point indicates the formation of ultimate clusters. A simple flowchart, given in Figure 3.5, gives an effective way to understand the algorithm at a glance.

The K-means algorithm has been used and adapted over the decades it has been in use, reaching an optimum form. Over the same period, many more data clustering algorithms, such as Fuzzy clustering (Yang, 1993), Fuzzy C-means (FCM) clustering (Bezdeck, et al., 1984), Expectation-maximization (EM) clustering (Mitchell, 1997), etc. have also been developed. But here, we have introduced the K-means algorithm as it meets our objectives regarding digital image watermarking.

CHAPTER 4

Copyright Protection Framework

I N THIS CHAPTER, WE develop a copyright protection tool for digital images through an image watermarking methodology. According to the basic concepts of digital watermarking discussed in Chapter 1, there are two basic operations in any watermarking system. One is watermark insertion and the other is watermark extraction. We then learned about the three basic properties of digital image watermarking and the contradictions between these. It was mentioned in Chapter 3 that utilization of certain intelligent techniques could be effective for improving all three properties up to a certain optimized level. Now, this discussion continues in two parts– watermark embedding and watermark extracting.

4.1 WATERMARK EMBEDDING SYSTEM

Watermark embedding or insertion is performed with a grayscale cover image and a binary watermark. It is better to understand the system functions properly with a single plane image compared with a color image, as we all know that any color image can be

FIGURE 4.1 Grayscale cover image.

decomposed into three basic single plane images (red, green, blue). A grayscale image C of size (M × N) has been chosen as a cover or host image (Figure 4.1). The fundamental theory of images says that C is basically an array of M × N positive integers, having values between 0 and 255. Mathematically, C can be defined as,

$$C = \{c(m,n)|1 \leq m \leq M,$$
$$1 \leq n \leq N \wedge c(m,n) \in [0,1,2,\ldots,255]\} \quad (4.1)$$

A binary image W of size (X × Y), shown in Figure 4.2, is taken as the watermark. Therefore, W will be an array of X × Y binary elements and defined as,

$$W = \{w(x,y)|1 \leq x \leq X, 1 \leq y \leq Y \wedge w(x,y) \in [0,1]\} \quad (4.2)$$

The watermark insertion is performed in the spatial domain, so that the execution time and computational cost will be less and the hardware implementation becomes easier. Its robustness should be a challenge, as it is a spatial domain technique with high data

FIGURE 4.2   Binary watermark.

transparency and payload. A step-by-step analysis of this data-embedding mechanism will show how the techniques described in Chapter 3 solve these issues.

Step 1: Evaluating the Saliency Map from the Cover Image

It has been determined that the purpose of a saliency map in image watermarking is to discover the regions with the least probability of changes to be perceived by the human visual system (HVS) and hence embed the maximum quantity of data into those portions, so that high imperceptibility can be achieved. Here, the spectral residual-based approach is used in generating the saliency map. As already explained, this method works on the principle of separating out the innovative or the novelty parts as salient objects and holding back the redundant components as prior knowledge. The spectral residual-based saliency map-generation process can be explained by reference to the block diagram in Figure 4.3 along with Eqs. (4.3) through (4.8) (Hou et al., 2007).

Now, the cover image C is considered to be the input image from which the saliency map is to be generated. In the spectral residual method, first, the frequency coefficients of the input image pixels are obtained through the discrete Fourier transform.

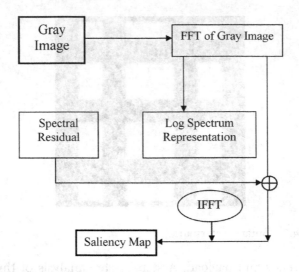

FIGURE 4.3 Block diagram for generating a saliency map of an image using the spectral residual approach.

$$C_f = \mathcal{F}[C] \tag{4.3}$$

where, $\mathcal{F}$ is the discrete Fourier transform operator.

From the Fourier spectrum, the arrays of amplitudes (C_A) and phases (C_P), having entities corresponding to each and every pixel of C, is determined as follows.

$$C_A = |C_f| \tag{4.4}$$

$$C_P = \angle C_f \tag{4.5}$$

The log spectra of C_A are estimated within the range of 0 to 1 and denoted by C_L.

$$C_L = \log_e(C_A) \tag{4.6}$$

Next, a local average filter is adopted to approximate the general shape of C_L. The filter, denoted by f_L, is a square matrix of order 'L' defined as,

$$f_L = \frac{1}{L^2} \begin{bmatrix} 1 & 1 & \cdots & 1 \\ 1 & 1 & \cdots & 1 \\ \vdots & \vdots & \vdots & \vdots \\ 1 & 1 & \cdots & 1 \end{bmatrix}$$

The general shape of the log spectra, defining the prior information, is obtained by convoluting f_L with C_L. The information from the smooth curve of log spectra C_L refer to the statistical singularities that are particular to C. These singularities are defined as the spectral residual (C_R) of the image C.

$$C_R = C_L - \left(C_L * f_L \right) \tag{4.7}$$

Therefore, C_R contains only the innovation of the input image. Using inverse Fourier transform to C_R, the saliency map S can be constructed in the spatial domain. The value at each point in a saliency map is then squared to indicate the estimation error. For better visual effects, the saliency map is smoothed with a Gaussian filter, f_g.

$$S = f_g * \left(\mathcal{F}^{-1} \left[e^{C_R + C_P} \right] \right)^2 \tag{4.8}$$

where, $\mathcal{F}^{-1}$ is denoted as the inverse discrete Fourier transform.

The saliency map of the cover image C is obtained through the spectral residual method as described in Eq. (4.8). The saliency map vector S can be defined as,

$$S = \left\{ s(m,n) \middle| 1 \le m \le M, \right.$$
$$\left. 1 \le n \le N, 0 \le s(m,n) \le 1 \wedge s(m,n) \in R \right\} \tag{4.9}$$

From Eqs. (4.8) and (4.9), it is clear that the saliency map is generated in such a manner so that the saliency of any pixel in the cover image can be referenced to its corresponding saliency map pixel value, and the saliency of the pixels increases with their values from 0 to 1. The saliency map for the cover image is shown in Figure 4.4. In this figure, brighter regions indicate higher saliency.

Step 2: Constructing the Hiding Capacity Map from the Saliency Map

The spatial domain watermarking schemes mostly follow the least significant bit (LSB) replacement technique to insert data into the cover image. Here also, the watermark bits are embedded by replacing one or more LSBs from each of the pixels of the cover image. In other words, an adaptive LSB replacement method is used in implanting the watermark bits into the cover image pixels. In a saliency map, the pixel values from 0 to 1 indicate the nature of the saliency of the corresponding pixel in the cover image in terms of the flat region to the busy region. The ability of the HVS to detect a slight change is less for the busy region compared with the flat region. For that reason, the pixels in the busy region are chosen to hide the maximum quantity of watermark bits to maintain the visual aesthetic. The watermarking scheme developed in this book allows

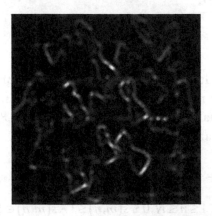

FIGURE 4.4   Saliency map for the input cover image shown in Figure 4.1.

replacing at least two LSBs from each and every cover image pixel with the watermark bits. The absolute difference between an original pixel value and its value after replacing the first two LSBs is only 3. Therefore, the visual distortions caused by this small distinction in pixel value will be minimal and will be unnoticeable to the HVS. For the pixels of the flattest or least busy regions, only the first two LSBs are to be replaced by the watermark bits. The number of replacing bits increases for the sectors having higher saliency. It will be noted that the hiding capacity increases with rising saliency because, in the spectral residual model, higher saliency indicates busier regions, having more edges or variations in neighboring pixel values. But we should keep in mind that for some other types of saliency map models, which detect the salient objects as the regions of interest according to the HVS (such as the Graph-Based Visual Saliency (GBVS) model or Itti's model), data hiding is to be decreased for higher saliency. Assume that the maximum number of LSBs that could be replaced in each pixel of the busiest region is chosen as K. If K is decreased by 1 with a falling in saliency, the cover image pixels should be divided into $K-1$ regions based on their saliency map values, as the minimum number of bits to be replaced is two. For this purpose, a hiding capacity map (HCM) has been developed, through which the cover image is separated into $K-1$ distinct regions. HCM generation is performed by fixing up several of the saliency map values. The K-means data clustering algorithm is employed here to group the cover image pixels into $K-1$ discrete regions according to the analogous saliency map values. This algorithm generates different centroids or means for each cluster or group. Therefore, we get $K-1$ centroids for $K-1$ regions. Equating all the pixel values of each cluster to its relevant centroid, the grouping can be visualized and the hiding capacity map $\widehat{H}$ is formed, as described in Eq. (4.10).

$$\widehat{H} = \left\{ \hat{h}(m,n) \middle| 1 \le m \le M, \right.$$

$$\left. 1 \le n \le N \wedge \hat{h}(m,n) \in \left[\sigma_1, \sigma_2, \ldots, \sigma_{K-1} \right] \right\}$$

(4.10)

where, σ_i indicates the centroid of the ith cluster, obtained through clustering over the saliency map S using the K-means algorithm. It is clear that the values of the centroids will be different for various images. To maintain uniformity in the watermark extracting system as well as in the hardware execution, the final HCM, denoted as H, is constructed as a function of $\hat{H}$. Now, if $h(m,n) \in H$ and its corresponding $\hat{h}(m,n) \in \hat{H}$, then the function $f_i: \hat{H} \rightarrow H$ is defined as,

$$h(m,n) = \left\lfloor \frac{255}{K-1} \right\rfloor (i-1) \quad \text{for} \quad \hat{h}(m,n) = \sigma_i \qquad (4.11)$$

where, $i \in [1, 2 \ldots K{-}1]$ and $i + 1$ is the number of LSBs chosen to be replaced, for each particular cover image pixel $c(m, n)$, by $h(m, n)$.

Therefore, the HCM is finally defined as,

$$H = \left\{ h(m,n) \middle| 1 \leq m \leq M, \right.$$
$$\left. 1 \leq n \leq N \wedge h(m,n) = f_i\big(\hat{h}(m,n)\big) \right\} \qquad (4.12)$$

The clustered output and the hiding capacity map for the cover image are shown in Figure 4.5a,b respectively.

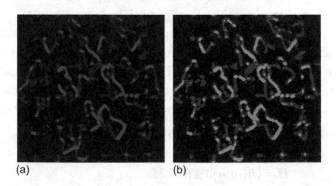

(a) (b)

FIGURE 4.5   (a) Clustered image; (b) hiding capacity map for the input cover image shown in Figure 4.1.

Step 3: Constituting the Watermarked Image through Adaptive LSB Replacement

The last stage of the embedding system is implanting the watermark bits through an adaptive LSB replacement technique, i.e. data embedding will be different for different pixels of the cover image. Moreover, this algorithm enables multiple embedding of the watermark. In this example, the size of the watermark is matched with the cover size by repeatedly placing the watermark image horizontally and vertically so that the size of the watermark is exactly the same as the size of the cover image. The new watermark, $W_{M,N}$, consists of multiple original watermarks W, and can be constructed as,

$$
\widehat{W} = \begin{bmatrix} W & W & \cdots & W \\ W & W & \cdots & W \\ \vdots & \vdots & \vdots & \vdots \\ W & W & \cdots & W \end{bmatrix}_{M \times N}
$$

Therefore, the new watermark $\hat{W}$ can be defined as,

$$
\widehat{W} = \left\{ w(m,n) \middle| 1 \leq m \leq M, 1 \leq n \leq N \wedge w(m,n) \in [0,1] \right\} \quad (4.13)
$$

The hiding capacity map is the key to navigating the adaptive bit replacement process. For any cover image pixel, a corresponding HCM pixel will be found. According to Eqs. (4.11) and (4.12), HCM pixels can have only a few discrete values, computed by

$$
\left(\left\lfloor \frac{255}{K-1} \right\rfloor (i-1) \right),
$$

for different values of 'i.' Depending on the values of 'i' for any HCM pixel, its analogous cover image pixel is modified by replacing its first '$i + 1$' LSBs by the resultant watermark bits. As the value of 'i' increases with the rise in saliency, data embedding also increases for higher salient regions.

If E is the watermarked image, $e(m,n) \in E$ and $w(m,n) \in \hat{W}$, then the function f_E: $C \times H \times \hat{W} \to E$ is defined as,

$$e(m,n) = \sum_{b=i+1}^{7} c_b(m,n) 2^b$$

$$+ \sum_{b=0}^{i} w(m,n) 2^b \quad \text{for} \quad h(m,n) = \left\lfloor \frac{255}{K-1} \right\rfloor (i-1)$$

(4.14)

where $c_b(m,n)$ is the bth bit (considering LSB as the 1st bit) of the pixel $c(m,n)$. Thus, the watermarked image E is finally defined as,

$$E = \left\{ e(m,n) \middle| 1 \le m \le M, 1 \le n \le N \wedge e(m,n) \right.$$

$$\left. = f_E \left(c(m,n), h(m,n), w(m,n) \right) \right\}$$

(4.15)

From Eq. (4.15) it is shown that the watermarked image E is a grayscale image with the same size as the cover image C. The final output of the embedding system, i.e. the watermarked image, is shown Figure 4.6.

In Figure 4.7, the complete embedding process has been summarized using a block diagram representation, specifying the steps involved in producing the watermarked image.

Hardly any difference could be perceived between Figures 4.1 and 4.6, which means that the original image and the watermarked image are identical to the HVS. This resemblance indicates the proficiency of this intelligent technique-based image watermarking scheme in terms of fine data transparency. However, as the images appear identical to the HVS, we could possibly question whether watermark is actually implanted into the original cover image through this watermarking methodology. To confirm this, another grayscale cover image of size 16×16 is taken (Figure 4.8a) and the proposed watermarking process is performed to generate a corresponding watermarked image by embedding a binary watermark (Figure 4.8b) of size 4×4 into this new cover image.

FIGURE 4.6 Watermarked image after embedding the watermark (shown in Figure 4.2), into the cover image shown in Figure 4.1.

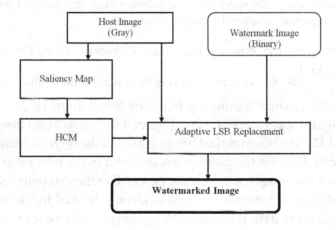

FIGURE 4.7 Block diagram for watermark embedding system.

The analogous matrices for the concerned images illustrate the purposes and outcomes of the steps followed in this watermark insertion procedure. The saliency map, clustered image, and the HCM, along with their corresponding matrices, are displayed in Figure 4.8c–e, respectively. Here we can see that the values of the saliency map image pixels vary within the range of 0 to 1. Choosing the maximum number of bits to be replaced as 4, the HCM is generated

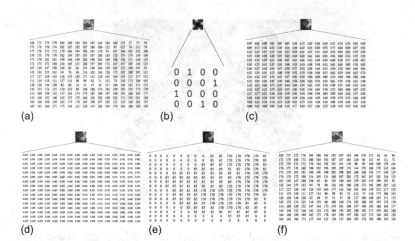

FIGURE 4.8 Illustration of watermark embedding with image arrays: (a) Cover image; (b) watermark; (c) saliency map; (d) clustered image; (e) hiding capacity map; (f) watermarked image.

by grouping the saliency map into 3 clusters. Thus, the value of $\left\lfloor \dfrac{255}{K-1} \right\rfloor$ is obtained as 85 and only three distinct values (0, 85 and 170), calculated using Eq. (4.11), are found in the HCM array. Finally, the watermarked image (Figure 4.8f) is developed through Eq. (4.15). The watermarked image is akin to the original image in this case also. But the difference is observed in the relevant arrays of those two images. It can clearly be seen that the maximum absolute difference between two pixels, identically sited in the original and watermarked image, is 15. Looking closely, we see that the analogous HCM pixel values are highest for the cover image pixels, for which this difference is obtained as 15. This difference trims down to 7 and 3 for the HCM values 85 and 0 respectively.

4.2 WATERMARK EXTRACTING SYSTEM

The purpose of watermark extracting is to recover the watermark from a received or watermarked image so that the originality can be evaluated by comparing it to the original mark. The received image can be defined as,

$$E_R = \left\{ e_R(m,n) \middle| 1 \le m \le M, \right.$$

$$\left. 1 \le n \le N \wedge e_R(m,n) \in [0,1,2,\ldots,255] \right\} \tag{4.16}$$

For the watermark extracting process, the same algorithm used for embedding the watermark is used for generating the HCM of the received image. Thus, for E_R, the analogous saliency map and HCM are obtained as S_R and H_R respectively. As the procedures to generate S_R and H_R are exactly similar to the processes discussed in embedding system, there is no need to repeat them here. Figure 4.9 demonstrates the watermark extracting system by a simple block diagram depiction.

The HCM of the watermarked image is used to obtain the number of watermark bits embedded into a particular cover image pixel. For any watermarked image pixel, if the corresponding HCM map value is

$$\left(\left\lfloor \frac{255}{K-1} \right\rfloor (i-1) \right),$$

then it indicates that '$i + 1$' bits are inserted in that particular pixel. Therefore, from that pixel, '$i + 1$' bits (starting from the first LSB) should be extracted. Signal processing attacks may distort the watermarked pixels during transmission. Thus, all the inserted watermark bits are extracted, although the same watermark bit is embedded multiple times for a particular cover image pixel. Finally, applying similarity estimation for the extracted bits from each pixel, the actual watermark bit is approximated and thus the probability of error in reconstructing the watermark is reduced. This similarity estimation process is the only exclusive step in this extracting system when compared with the embedding system. This operation is described below.

Let the bits, extracted from any particular watermarked image pixel, be noted as $b_0, b_1, \ldots, b_i$, considering b_0 as the first LSB.

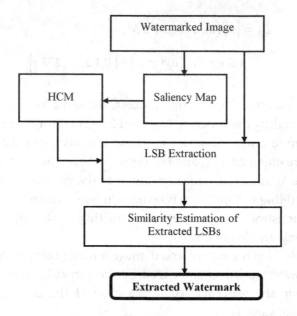

FIGURE 4.9   Block diagram for watermark extracting system.

These extracted bits can be thought of as a small set of binary numbers, i.e. the set of the extracted watermark bits for each received image pixel consists of binary elements having the values 0 or 1. A single bit is approximated from each of the sets through similarity estimation. If the approximated watermark bit is w_r, then the similarity estimation function assigns the value of w_r as 0 when the occurrence of 0 within that particular set is not less than $\left\lceil \dfrac{i}{2} \right\rceil$. In the same way, the value of w_r is assigned to 1 for the reverse case. If the occurrence of 0 and 1 is same, w_r could be either 0 or 1. Now, if the similarity estimating function is noted as f_b and $B_{m,n}$ is considered to be a set of the bits $(b_0, b_1, \ldots, b_i)$, extracted from any particular watermarked pixel $e_R(m,n)$ then, $w_r = f_b(B_{m,n})$, where the construction of $B_{m,n}$ for each received pixel $e_R(m,n) \in E_R$ depends on that pixel value along with

FIGURE 4.10 Extracted watermark from the watermarked image shown in figure 4.6.

its corresponding HCM value $h_R(m,n) \in H_R$. In other words, $B_{m,n}$ is obtained as a function of $e_R(m,n)$ and $h_R(m,n)$, i.e. $B_{m,n} = f(e_R(m,n), h_R(m,n))$.

Now, we should remember that the watermark was inserted multiple times to produce the watermarked image by repeating the watermark pattern at the time of embedding. Hence, after considering all X × Y pixels, the extraction of one watermark is executed. Thus, the recovered watermark is finally obtained as a function of E_R and H_R and can be stated as,

$$W_R = \left\{ w_R(x,y) \middle| 1 \le x \le X, \right.$$
$$\left. 1 \le y \le Y \wedge w_R(x,y) = f_b(B_{x,y}) \right\} \tag{4.17}$$

where,

$$B_{x,y} = f\left(e_R(x,y), h_R(x,y)\right) \wedge e_R(x,y) \in E_R, h_R(x,y) \in H_R$$

If we consider the image shown in Figure 4.6, to be the received or watermarked image, the watermark extracted from it through this proposed method and the recovered watermark is shown in Figure 4.10. Readers can find that the recovered watermark in Figure 4.9 is identical to the original watermark, given in Figure 4.2. Therefore, we can conclude that the watermark embedding and extracting systems developed in this chapter are working properly. The next chapter will lead us to execute the hardware implementation of this proposed watermarking methodology.

CHAPTER 5

Hardware Implementation

REAL-TIME EXECUTION IS REQUIRED for any new algorithm in order to enhance its efficiency in terms of speed, space, and cost. Progress in large-scale system designing and digital integrated circuit technologies results in incomparable growth in electronics industries. Very-large-scale integration (VLSI) systems are capable of implanting up to ten million transistors in a single chip to solve up to twenty thousand logic complexities (Sarkar et al., 2014). VLSI design methodologies are broadly divided into two types: full-custom and semi-custom (Kang and Leblebici, 2003). Full-custom designs, such as application-specific integrated circuits (ASIC), are capable of providing superior proficiency in terms of processing speed, area utilization, and power dissipation. But the high computational cost, caused by greater design time, is an alarming issue for full-custom VLSI designs. On the contrary, semi-custom designs, such as field programmable gate arrays (FPGA) or standard cell-based designs, require less time to execute the operational functions, and the system performance can also be optimized. The solid performance of high-definition applications of recent FPGA is a good reason to prefer it to the hardware implementation of software-based algorithms. Current FPGA chips are capable of running graphics algorithms at a speed comparable with that of dedicated graphics chips. Furthermore, these FPGA-based designs are simple to implement, providing good flexibility.

5.1 OVERVIEW OF HARDWARE REALIZATION FOR DIGITAL WATERMARKING

During the last few years, hardware-driven digital watermarking logic, developed in either the spatial or the frequency domains, is being performed based on FPGA or other VLSI-design models. P. Zemcik explains that computer graphics algorithms and image processing algorithms are, in general, computationally expensive. This is why people struggle to develop such algorithms using reasonable means, such as faster processors, parallelism, or dedicated hardware (Zemcik, 2002). These algorithms are configurable using schematics diagrams and also through high-level programming languages, for example, VHDL (VHSIC hardware description language, where VHSIC means very high speed integrated circuit). Section 5.1 addresses these issues as well as the general developments in this field, and gives examples of hardware platforms and algorithms that can be implemented on such platforms.

An FPGA-based implementation of blind and invisible watermarking on Altera FPGA was presented by Seo and Kim (Seo and Kim, 2003). The algorithm was realized in the discrete cosine transform (DCT) domain and the discrete cosine (DC) coefficients were replaced by watermarking, which was imperceptible to the human eye. The watermarking algorithm was incorporated using a JPEG 2000 encoder at an operating frequency of 66 MHz. In 2004, Mohanty, Kumara, and Nayak described an invisible, robust spatial domain watermarking algorithm and its FPGA implementation (Mohanty et al., 2004). The algorithm is non-blind, and watermark insertion is carried out by replacing the original image pixel value by a watermark encoding function. The scheme was evaluated using a standard benchmark, such as StirMark software, and realized on an XCV50-BG256-6 device from Xilinx, where the operating frequency was 50.398 MHz.

The VLSI architecture of biometric-based watermarking was described by S. P. Mohanty et al. (2006). The algorithm works for both grayscale and color images, which act as host images, with a biometric image being selected for the watermark. The original image is divided into 8 × 8 blocks and the DCT is calculated for each block. The watermark is divided into blocks and embedded into a perceptually significant region of the host image. This non-blind approach makes the watermark robust against general signal processing attacks. The architecture was modeled using VHDL and implemented on an XC2V500-6FG256 Xilinx device. In 2007, SysGen-based architecture for filtering images was described by Sánchez et al. (2007). This design helped to enhance image characteristics by eliminating noise, enhancing edges and contours, etc. It focused on the processing of pixel to pixel of an image and on the modification of pixel neighborhoods; and, of course, the transformation could be applied to the whole image or only a part of it. Wagh et al. described the implementation of the JPEG 2000 Encoder using VHDL (Wagh et al., 2008). The proposed architecture used the lifting-scheme technique and provided advantages that included low memory requirements, fixed-point arithmetic implementation, and a small number of arithmetic computations. The architecture was modeled using VHDL, and a function simulation was performed. This chip was tested using AccelDSP in a hardware-in-the-loop (HIL) arrangement. The proposed scheme was robust against several geometric attacks. Saraju P. Mohanty (2009) projected an innovative non-blind algorithm for encrypted watermarking based on block-wise DCT. The algorithm could process grayscale images as well as color images as the host images. In the case of color images, the host image in RGB format was converted into YCbCr (which is a popular color space in computing), with the Y component being selected for watermarking. The image is divided into 8 × 8 blocks and the DCT was calculated for each block. The encrypted watermark was embedded into the transformed image using four different embedding strength factors, chosen such that the image quality would not degrade. The block-wise DCT was calculated for both images and the difference calculated in order to detect a watermark. The extracted watermark was then compared with the original watermark for authentication.

Human visual system (HVS)-based image adaptive watermarking and its hardware architecture were described by Lande et al. (2009). The proposed scheme of watermarking was invisible and robust against JPEG attacks. The host image was divided into 8 × 8 blocks and the discrete Hadamard transform (DHT) calculated. A pseudo-random noise (PN) sequence was generated through a user key and embedded into the DHT coefficients. The strength factor was calculated from the quantization table for the DHT domain. The proposed method was blind and robust against common signal processing attacks, such as low-pass filtering and noise addition. The algorithm was implemented using an XC3SD1800A-4FGG676C Xilinx device and functional simulation was performed using Xilinx tools. The implementation was verified using hardware co-simulation at 33.3 MHz. Another HVS-based digital watermarking technique was developed by A. Basu et al. (2015) where software-hardware co-simulation was employed. This scheme fully utilized the imperfect nature of the human visual system and injected information adaptively into the cover so that the maximum amount of data could be hidden in the least interested regions. The FPGA execution of that logic was performed through an XC7A30T-3CSG324 Xilinx device and the language was assigned VHDL. This method enabled a system to embed a watermark in a 64 × 64 grayscale cover image, and to recover the mark from the watermarked image in 1750 nanoseconds, considering the transmission time to be zero.

Que et al. presented the process of implementing a full system to reconstruct CT (computed tomography) images on FPGA using Simulink and SysGen (Que et al., 2010). A cone-beam back projection system under the Feldkamp-Kress-Davis (FDK) algorithm using Simulink and SysGen was implemented. When the image was sampled, the samples of the image were very concentrated toward the center and very sparse near the edges. To compensate such a mismatch, a filter needs to be applied to the projection data. A video watermark technique was proposed by El-Araby et al. (2010). The technique depended on embedding an invisible watermark in a low-frequency DCT domain by means of a pseudo-random number sequence generator for the video frames in place of high- or mid-band frequency components. This method was realized using MATLAB® and VHDL, and the system was implemented on an XC5VLX330T Xilinx device. The result of implementation showed that the maximum frequency for real-time operation is 13.61 MHZ.

An invisible watermarking algorithm for color images was developed by T.J. Anumol et al. (2013) where the RGB image was converted to HSV first, and then data embedding was performed using discrete wavelet transform (DWT). A maximum frequency of 344 MHz is offered by its FPGA implementation through a 6vsx315tff1156-2 device. In 2014, a DCT-based robust image watermarking scheme and its FPGA design were described by R.M. Khoshki et al. (2014) where the Alterra DSP (digital signal processing) Builder was integrated with the Simulink embedded coder in order to generate the secret code automatically. H. K. Maity and S. P. Maity introduced reversible contrast mapping (RCM)-based reversible watermarking (Maity and Maity, 2014) and, with its FPGA implementation, achieved an improved operation frequency of 98.76 MHz. To operate at this high frequency with a data rate of 1.0395 Mbps, this architecture involved a six-stage pipeline technique with 9,881 slices, 9,347 slice flip-flops, 11,291 4-input Lookup Tables (LUT), and 3 Block RAMs (BRAM).

A novel transform domain digital image watermarking using discrete fast Walsh–Hadamard transform was formulated by S. Ghosh et al. (2014), and the hardware realization developed was based on FPGA. Here xc7vx1140t-1flg1930 was chosen as the target device to achieve a maximum frequency of 259.202 MHz. This method has been used for both grayscale and binary watermarks. A recent work of M. R. Nayak et al. (2017) defines an effortless watermarking algorithm based on phase congruency and the singular value decomposition technique. Its hardware realization up to the logic schematic level is carried out using an FPGA board, consuming a low dynamic power of 5.029 mW with only 1.539 ns delay; after each clock pulse, it delays for the given time to carry out a particular operation.

In the next section, we are going to look at the architecture of the hardware systems for the watermark embedding and extracting algorithms (as described in the previous chapter). Following this, the FPGA implementation of these hardware architectures is discussed. Being developed in the spatial domain, the hardware architecture for this watermarking scheme can be developed with less complexity and a lower computational cost. The system architectures as well as system operations are described through some well-known combinational and sequential digital circuit elements, for example, logic gates, multiplexers, registers, counters, etc. (Mano and Ciletti, 2013; Salivahanan and Arivazhagan, 2007).

5.2 HARDWARE ARCHITECTURE FOR THE PROPOSED IMAGE WATERMARKING SCHEME

5.2.1 Watermark Embedding System

The watermark embedding system proposed in Chapter 4 adaptively implants a watermark into the cover image pixels based on the corresponding hiding capacity map (HCM) pixel values and produces the watermarked image pixels as the system output. As one would expect, the watermark bits and the cover image pixels, along with the analogous HCM pixels, are considered to be the inputs of this watermark embedding system. The clock (CLK) input is taken to synchronize the embedding operation and the reset (RST) input is used to reset the states of the system components for distinct operations. As the cover image is a grayscale image, each of the cover image pixels consists of 8 bits. Similarly, every HCM pixel also consists of 8 bits. Thus, two 8-bit shift parallel input parallel output (PIPO) registers are taken as input registers and another 8-bit shift PIPO register is used to store modified pixel bits, which generate the system output (i.e. the watermarked image pixels). Figure 5.1 shows the operational blocks of hardware architecture of this embedding system.

The steps that are then followed in the hardware implementation of the proposed embedding method are presented here, based on a maximum 4 bits to be replaced.

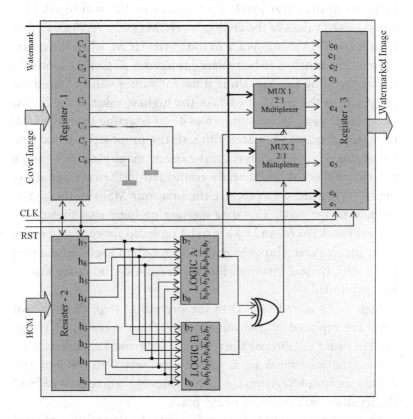

FIGURE 5.1 Hardware architecture for the watermark embedding system.

Step 1: The cover image pixels are taken one at a time as an input and stored in Register-1. At the same time, the analogous HCM pixel is supplied to Register-2. The watermark bits are taken one by one as another input of the embedding system. Let, c_0, c_1, ..., c_7 be the bits of the cover image pixel taken at any instant of time, and the corresponding HCM pixel bits are noted as h_0, h_1, ..., h_7. Here, c_0 and h_0 indicate the LSBs of the concerned image pixels, whereas, c_7 and h_7 specify the most significant bits (MSB). The watermark bit provided to the input at that particular instant of time is designated as w.

Step 2: According to the proposed data embedding logic, multiple bits in the cover pixels are replaced by the watermark bits, based on the values of the analogous HCM pixels. As discussed in Chapter 4, $K-1$ variations are found in the HCM, when the maximum number of bits to be replaced is fixed at K. Consequently, in this embedding process, three different values will be found for the pixels in the HCM, if we take the highest value of the bits to be replaced as 4. As a maximum of 4 bits (starting from the LSB, i.e. c_0 to c_3) may be replaced through the proposed method, the remaining 4 bits (i.e. c_4 to c_7) of the cover image pixel will always remain unaffected and directly connected to the corresponding inputs of Register-3 to produce the first four MSBs of the watermarked image pixel. This data implanting logic also shows that at least two LSBs (c_0 and c_1) are to be replaced for each and every cover image pixel. Thus, the embedding system itself has nothing to do with these 2 bits and, for this reason, c_0 and c_1 are always kept grounded.

Step 3: As already stated in the preceding step, the first two LSBs are replaced by the watermark bit for every cover image pixel in order to fabricate the resultant watermarked image pixel. Hence, the watermark bit is directly connected to the first and second inputs of Register-3, which produces the first two LSBs of the resultant watermarked image pixel.

Step 4: From Eq. (4.11), we can compute the three different pixel values of HCM as 0, 85, and 170. These values ascertain the number of bits (starting from the LSB) to be replaced for the analogous cover pixel as 2, 3, and 4 respectively. Bit replacement for the first two LSBs is performed in the required way, as described in Step 3 and, incidentally, this also solves the bit replacement issues for the condition when the HCM pixel value is 0. Now, the bit replacement for the third and fourth LSBs (c_2 and c_3) is a conditional process. Two 2:1 multiplexers (MUX-1 and MUX-2) are present in this framework. c_2 is connected to the first input of MUX-1 and c_3 is connected to the first input of MUX-2. The watermark is considered as the second input for both of the multiplexers. The outputs of MUX-1 and MUX-2 are linked up to the third (e_2) and fourth (e_3) inputs of Register-3 respectively. Each of the multiplexers transfers its first input to output with the select value 0 and the output equal to the second input for the select value 1. In other words, the select line values of MUX-1 and MUX-2 determine whether the third and fourth bits will be replaced or not. For the select value 1, the relevant bit is replaced by the watermark bit, otherwise, not. The select line values depend on the outputs of the two logic blocks, shown in Figure 5.1 as LOGIC-A and LOGIC-B.

Step 5: The logic blocks act as a control unit, influencing the multiplexers to set the values of their select lines. Both of the logic blocks take HCM pixel bits as inputs, noted as b_0, b_1, ..., b_7, respectively related to h_0, h_1, ..., h_7. LOGIC-A block describes its output function as $\overline{b_0}b_1\overline{b_2}b_3\overline{b_4}b_5\overline{b_6}b_7$ and the output function of LOGIC-B is described as $\overline{b_0}b_1\overline{b_2}b_3\overline{b_4}\,\overline{b_5}\,\overline{b_6}\,\overline{b_7}$.

The LOGIC-A output is 1 if and only if the HCM pixel value is equal to 170. This logic output connects with the select line of MUX-2. Now, the second input of MUX-2, i.e. the watermark bit, is sent to the output with select line value 1. Hence, the fourth LSB (c_3) of the cover image pixel is replaced by the watermark bit w. The LOGIC-B output is set to 1 only for the HCM pixel value 85. The first three LSBs are replaced by the watermark bit for this condition. But the third bit is also to be replaced for the HCM pixel value 170. Thus, the output of LOGIC-B is XORed with LOGIC-A output and the XORed output contributes to the select line values of MUX-1.

As a result, the select line value is 1 for either of the outputs of the logic blocks and bit replacement is performed for the third LSB. When both of the logic blocks generate 0 as output, the XOR operation provides 0 to the MUX-1 select line and the original pixel bit is transferred to the MUX output, i.e. no bit replacement occurs.

Step 6: Finally, Register-3 constructs the embedded or watermarked image pixel, such that the first four MSBs (i.e. e_7 to e_4) are exactly identical to the parallel original bits (i.e. c_7 to c_4), the first two LSBs (e_0 and e_1) are always set to the instantaneous watermark bit w and the remaining 2 bits (e_2 and e_3) are conditionally replaced according to the output of the logic blocks.

In this way, the hardware implementation of the watermark embedding system is achieved to generate the watermarked image pixel for an original image pixel through the proposed intelligent technique-based adaptive bit replacement method, with the watermark bit immediately supplied to the input.

5.2.2 Watermark Extracting System

The watermark extracting technique is basically the reverse of the watermark embedding technique. The hardware architecture of proposed watermark extracting system is shown in Figure 5.2.

In case of watermark extraction, the pixels of a watermarked image are considered to be the system input and stored into Register-1 (one pixel at a particular time instant). Akin to the embedding system, the corresponding HCM pixel for that watermarked pixel is taken to Register-2 as another input. The retrieved watermark bits from consecutive pixels of the watermarked image are accumulated in Register-3 to reconstruct the watermark as hidden in that particular watermarked image. Here, Register-1 and Register-2 are 8-bit PIPO shift registers and Register-3 is an n-bit serial input serial output (SISO) register where n is the number of bits present in the watermark sequence.

The steps performed to achieve the function of the watermark extracting system, relative to the data embedding methodology used, are described below.

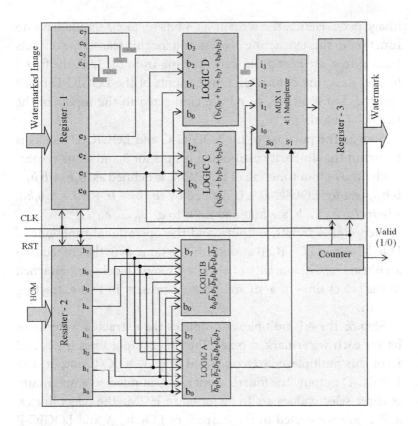

FIGURE 5.2   Hardware architecture for the watermark extracting system.

Step 1: At successive instants of time, the consecutive pixels of the watermarked image along with their corresponding HCM pixels are respectively taken to the inputs of two input registers, such that, at any particular time, Register-1 consists of a single watermarked pixel and its analogous HCM pixel is present in Register-2. For every watermarked image pixel, Register-1 generates 8 outputs, noted as e_0, e_1, ..., e_7, and h_0, h_1, ..., h_7 are the outputs from Register-2. Similar to the embedding architecture, here also e_0 and h_0 stand for the LSBs and e_7 and h_7 indicate the MSBs.

Step 2: According to the embedding logic, watermark bits may exist only in the first 4 LSBs (e_0, e_1, e_2, and e_3) of any watermarked image pixel. Hence, the remaining 4 bits (e_4 to e_7) should have no function in this extracting process and for this reason the outputs from e_4 to e_7 are kept grounded. Starting from the LSB, the first 4 bits (e_0 to e_3) are connected to the inputs of the LOGIC-D block and the first 4 bits (e_0 to e_2) are connected to the inputs of the LOGIC-C block.

Step 3: The purpose of the LOGIC-C and LOGIC-D blocks is to obtain the similarity estimation results for 3 and 4 bits respectively. The output function of LOGIC-C is defined as ($b_0 b_1 + b_1 b_2 + b_2 b_0$), and for LOGIC-D it is defined by ($b_3 (b_0 + b_1 + b_2) + b_0 b_1 b_2$), where, b_0, b_1, ..., b_3 are homogeneous to e_0, e_1, ..., e_3.

Step 4: The inputs, outputs, and the operational functions of LOGIC-A and LOGIC-B are exactly the same as in the embedding architecture. The outputs of these two logic blocks are attached to the select lines of a 4:1 multiplexer, present in this extracting system.

Step 5: The 4:1 multiplexer produces the extracted watermark bit for each watermarked pixel. The first 3 input lines (i_0, i_1, and i_2) of this multiplexer are connected to e_1, LOGIC-C output, and LOGIC-D output. The fourth input i_3 is grounded, as a maximum of three select values are to be found in HCM. The select lines s_1 and s_0 are connected to the outputs of LOGIC-A and LOGIC-B respectively. When both of the select line values are 0, it indicates neither of the outputs of LOGIC-A and LOGIC-B is 1, that means the HCM value is 0 and hence e_0 or e_1 could be the watermark bit. Thus, for the condition, $s_0 = 0$ and $s_1 = 0$, i_0 is transmitted to the MUX output. Similarity estimation for only 2 bits is futile and thus the second LSB is connected to i_0 as the first LSB has the highest error probability. When $s_0 = 1$ and $s_1 = 0$, it specifies that the LOGIC-B output is 1 and LOGIC-A output is 0. This situation occurs when the HCM pixel as the value 85. This value shows that any of the first three LSBs could be the watermark bit, and that this could be confirmed through LOGIC-C. Thus, this combination of select values sets the MUX output at i_1. In the case of the reverse condition, i.e. $s_0 = 0$ and $s_1 = 1$, it is simply asserted that the

HCM pixel value is 170 and the watermark bit may exist in the first four LSBs of the input image pixel. In that case, LOGIC-D should be in use and that's why the third input (i_2) of the multiplexer is considered as the MUX output as well as the extracted watermark bit. $s_0 = 1$ and $s_1 = 1$ is an invalid condition as both of the outputs of LOGIC-A and LOGIC-B cannot be 1 for a particular HCM pixel. This way, every watermark bit is individually retrieved from every watermarked image pixel successively supplied to the extracting system input.

Step 6: A counter is present in this extracting block. The counter value, initiated at 0, is increased by 1 after retrieving each bit. These steps above are repeated until the counter value is equal to the size of the watermark sequence. As soon as the counter value is equal to the size of the watermark, it indicates the completion of a watermark extraction process and the set of retrieved bits in Register-3 is considered as the watermark. Then the counter value is set to 0 again and from the next watermarked image pixel, another extracting process is initiated. In this way, multiple numbers of watermarks can be retrieved to compare to the original watermark to verify data authenticity.

The hardware execution of the proposed watermark embedding and extracting methodologies is realized through FPGA designs. The reason of choosing FPGA as the design mode is discussed at the beginning of this chapter. Most of the large FPGA-based systems are formed through complicated processes, consisting of several complex transformations and optimizations. Therefore, software tools are used to automate this type of system. Toward this end, Xilinx (ISE version 13.2) has been used to implement and synthesize the FPGA-based hardware systems for both of the watermark embedding and extracting algorithms. The experimental analysis for this FPGA implementations is established in the next chapter (Section 6.6), where the register-transfer level (RTL) schematics and corresponding simulation results attest the accuracy of the system.

CHAPTER 6

System Evaluation

I N CHAPTER 4, AN intelligent technique-based image water-marking scheme was developed for the purpose of copyright protection for digital images. The field programmable gate arrays (FPGA) implementation of this proposed method was then synthesized and verified in Chapter 5. We now need to weigh up the system proficiency, so that the acceptability of this proposed watermarking methodology can be assessed. In Chapter 1, we found that payload capacity, imperceptibility, and robustness are the three major parameters used to judge the efficiency of any image watermarking scheme. In this chapter, the system performance is judged on these three parameters and compared with some existing watermarking frameworks. Later, the hardware system is also evaluated in terms of computation speed, complexity, and utilization of logic components. Initially, our discussion begins with the embedding system outcomes, from which payload capacity and imperceptibility can be assessed.

6.1 EMBEDDING SYSTEM OUTCOMES

In Chapter 4, for a grayscale cover image and binary watermark image, the output result of the embedding system, i.e. the watermarked image together with the outputs of the intermediate functions (e.g. saliency map, clustered image, and hiding capacity map (HCM)) of the embedding block, is displayed to show how the original image is gradually modified with embedding operations and the ultimate output or the watermarked image is shaped. For the evaluation of this data embedding technique, selected images from the USC-SIPI image database, broadly used in the domain of digital watermarking as well as image processing, have been employed as cover images to undergo the proposed watermarking process. Each of the database images is adjusted to be a 256 × 256 grayscale image, before applying it to the embedding system input as the cover or host image. A binary image of size 64 × 64 is produced as the watermark, as shown in Figure 6.1a. There are more than 40 images in the USC-SIPI image database. Thus, we have given the output results of the embedding function only for

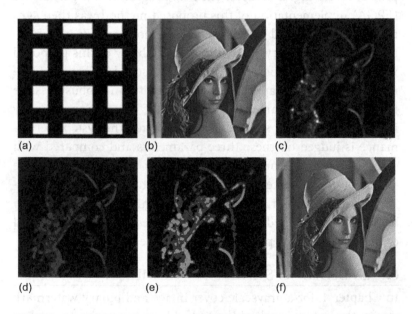

(a) (b) (c)

(d) (e) (f)

FIGURE 6.1   (a) Binary watermark (64 × 64); (b) grayscale cover image (256 × 256); (c) saliency map; (d) clustered image; (e) hiding capacity map; (f) watermarked image.

a single image from that database (i.e. 4.2.04.tiff, which is one of the most popular images in the watermarking domain and is known as the Lena image), although the watermark insertion process was applied to all the database images in order to compute and assess the output results. As we have already learned, the watermarked image is obtained as the absolute outcome of the watermark embedding system, but, for this proposed methodology, saliency map, clustered image, and hiding capacity map are generated as intermediary results. These are all connected to the original cover image and have their own individual contributions in forming the final watermarked image. The original image (with its corresponding saliency map, clustered image, and HCM) and the watermarked image are shown in Figure 6.1.

In Figure 6.1c, it can be seen that the brightness of the saliency map pixels increase in busy regions, where changes to the neighboring pixel values occur very frequently. The human visual system, by its nature, is unable or less able to detect slight changes in these busy regions. Thus, maximum data can be implanted in busy regions and less data in the flat regions. The proposed watermarking system performs watermark insertion on the basis of adaptive bit replacement and, as already discussed, the saliency map pixels are grouped in $K-1$ clusters for replacing maximum K bits. We have applied the proposed watermarking logic to develop a system that can replace a maximum of four bits from a pixel. The saliency map pixels are classified into three regions in the clustered image (Figure 6.1d). The hiding capacity map is developed to offer generic pixel values to the clustered image pixels, as well as showing where the visual distinctions between the pixels of different regions are more prominent in HCM, compared to the clustered image. In this hiding capacity map, shown in Figure 6.1e, the brightest areas indicate the most favorable regions for data embedding and the duller regions indicate areas where there is less hiding capacity. Finally, the watermarked image is exhibited in Figure 6.1f. It is clear from Figure 6.1f that the watermark remains perceptually transparent in the watermarked image, as hardly any visual distortion is found in the watermarked image with respect to the original image.

6.2 ESTIMATION OF DATA HIDING CAPACITY

An important feature of a good watermarking scheme is the data hiding capacity or payload capacity. It is measured in bits per pixel (bpp) and defined as the average number of watermark bits inserted into each pixel of the host image. Mathematically,

Data hiding capacity

$$= \frac{\text{Total number of bits implanted into cover image}}{\text{Total number of cover image pixels}} \text{bpp} \quad (6.1)$$

The formulation of the payload capacity depends on the watermark embedding logic and it may vary with different algorithms. For the watermarking scheme developed in this book, the maximum data hiding capacity can be calculated as,

$$\text{Maximum data hiding capacity} = \frac{\sum_{i=1}^{K-1} P_i B_i}{M \times N} \text{bpp} \quad (6.2)$$

where, P_i indicates the number of pixels in the ith cluster and B_i is the number of bits to be replaced for each pixel in that ith cluster. As defined before, K is the maximum number of bits to be replaced. Selecting the value of K as 4, this proposed algorithm is applied for all the images of the USC-SIPI image database, and the average hiding capacity is estimated as 2.30 bits/pixel. This payload is sufficiently good for a spatial domain watermarking scheme. Later in Section 6.5, we can see that our method is able to surpass many of the existing watermarking schemes in term of payload, data embedding, or data hiding capacity.

6.3 ANALYSIS OF IMPERCEPTIBILITY OR DATA TRANSPARENCY

6.3.1 Description of Image Quality Metrics, Involved in Imperceptibility Analysis

Imperceptibility in digital watermarking stands for the visual transparency of the watermark data embedded into the cover image. In other words, a watermarking system will be considered to provide good imperceptibility if it is capable of implanting data into an original image with minimal visual degradation. Computing the visual discrepancies between the original image and the corresponding watermarked image, data transparency can be evaluated. To this end, a set of image quality metrics have been produced. Quality metrics (Kutter and Petitcolas, 1999; Hore and Ziou, 2010) may be defined either as some comprehensive measures, in quantitative or qualitative form, that can be used to assess the quality of an adapted image on the basis of the original one (full reference), or some random scores for an image regardless of any original reference (no reference). Image quality metrics play the central role in depiction of many visual signal processing algorithms, such as imaging, cryptography, watermarking, etc. In digital watermarking, these metrics are very important for differentiating between the original and the modified images, so that system performance can be verified and improved up to a certain level of optimization. In this section, we have utilized several full-reference-based image quality metrics to assess imperceptibility by computing the differences, distortions, or correlations between the original cover image and the watermarked image.

Signal-to-noise ratio (SNR) is one of the most popular quality metrics, widely used not only for images but also for other types of signals. Usually, it is defined as the ratio of signal power to noise power. Considering an original image of size M × N as C and its analogous noisy image as E, SNR could be formulated as,

$$\text{SNR} = 10\log_{10}\frac{\sum_{m=1}^{M}\sum_{n=1}^{N}\left[c(m,n)\right]^2}{\sum_{m=1}^{M}\sum_{n=1}^{N}\left[c(m,n)-e(m,n)\right]^2}dB \quad (6.3)$$

where, $c(m,n)$ is any cover image pixel, such that $c(m,n) \in C$ and $e(m,n)$ is its corresponding watermarked image pixel, satisfying the condition $e(m,n) \in E$. For convenience, this notation (i.e. C as cover image, E as watermarked image, $c(m,n)$ and $e(m,n)$ for image pixels belonging to C and E respectively, and M × N as image dimension) are used in defining quality metrics throughout this section. A higher value of SNR indicates lower noise, which results in better image quality as well as enhanced data transparency.

Dividing the noise power by the total number of image pixels, the mean square error (MSE) is used as another image quality metrics and is defined by,

$$\text{MSE} = \frac{1}{MN}\sum\nolimits_{m=1}^{M}\sum\nolimits_{n=1}^{N}\left[c(m,n)-e(m,n)\right]^{2} \qquad (6.4)$$

It is clear from the above equation that the lower values of MSE cause less distortion in the pixels.

Substituting the numerator and denominator of the SNR equation (Eq. (6.3)) respectively by the maximum signal power and MSE, the peak signal-to-noise ratio (PSNR) is computed as,

$$\text{PSNR} = 10\log_{10}\frac{\sum\nolimits_{m=1}^{M}\sum\nolimits_{n=1}^{N}\left[c(m,n)\right]_{\max}^{2}}{\text{MSE}}dB \qquad (6.5)$$

Like SNR, high PSNR values indicate high imperceptibility. Typically, a 40 dB PSNR value is an excellent value for any digital watermarking scheme.

The maximum difference (MD) between the image pixels of C and their analogous pixels in E is computed by analyzing the absolute differences for all image pixel pairs, i.e. $\left|c(m,n)-e(m,n)\right|$ for all values of m and n, where, $m \in [1, 2, \ldots M]$ and $n \in [1, 2, \ldots N]$.

Obtaining the average absolute difference (AD) for the image pixels will be more effective in defining the presence of distortions throughout the overall image. It is defined as,

$$AD = \frac{1}{MN} \sum_{m=1}^{M} \sum_{n=1}^{N} |c(m,n) - e(m,n)| \qquad (6.6)$$

It is expected that lower AD as well as MD will indicate better data transparency.

Two further image quality metrics, the normalized average absolute difference (NAD) and the normalized MSE (NMSE), are formed by normalizing the values of AD and MSE respectively. These metrics are mathematically described as,

$$NAD = \frac{\sum_{m=1}^{M} \sum_{n=1}^{N} |c(m,n) - e(m,n)|}{\sum_{m=1}^{M} \sum_{n=1}^{N} c(m,n)} \qquad (6.7)$$

$$NMSE = \frac{\sum_{m=1}^{M} \sum_{n=1}^{N} (c(m,n) - e(m,n))^2}{\sum_{m=1}^{M} \sum_{n=1}^{N} (c(m,n))^2} \qquad (6.8)$$

Naturally, for two identical images, the values of NAD and NMSE will be found to be 0. That is why, to obtain higher imperceptibility, the values of these two metrics should be set to zero.

The deviation between unity and the NMSE for an image is termed image fidelity (IF) for that particular image. IF can be written as,

$$IF = 1 - \frac{\sum_{m=1}^{M} \sum_{n=1}^{N} (c(m,n) - e(m,n))^2}{\sum_{m=1}^{M} \sum_{n=1}^{N} (c(m,n))^2} \qquad (6.9)$$

From the above equation, it is clear that the IF value lies between 0 and 1 for any pair of images and a higher value of IF indicates improved data transparency.

The Laplacian mean square error (LMSE) is also used in imperceptibility assessment. A lower LMSE value (0 for equal image signals) indicates better imperceptibility. It is calculated by the following equation.

$$\text{LMSE} = \frac{\sum_{m=1}^{M}\sum_{n=1}^{N}\left(\nabla^2 c(m,n) - \nabla^2 e(m,n)\right)^2}{\sum_{m=1}^{M}\sum_{n=1}^{N}\left(\nabla^2 c(m,n)\right)^2} \quad (6.10)$$

where, for any arbitrary image pixel $i(m,n)$,

$$\nabla^2 i(m,n) = \sum \left(i(m+1,n), i(m-1,n), i(m,n+1), i(m,n-1)\right)$$
$$-4 \cdot i(m,n)$$

The noise quality measure (NQM) is estimated by,

$$\text{NQM} = 10\log_{10}\frac{\sum_{m=1}^{M}\sum_{n=1}^{N} e(m,n)}{\sum_{m=1}^{M}\sum_{n=1}^{N}\left|e(m,n) - c(m,n)\right|}dB \quad (6.11)$$

The watermark to document ratio (WDR) could be another quality metric, defined by the ratio of watermark energy to the cover image energy. Therefore,

$$\text{WDR} = 10\log_{10}\frac{\sum_{m=1}^{M}\sum_{n=1}^{N}\left(c(m,n) - e(m,n)\right)^2}{\sum_{m=1}^{M}\sum_{n=1}^{N} c(m,n)^2}dB \quad (6.12)$$

All the quality metrics for visual data analysis discussed so far are categorized as difference distortion measures (Sayood, 1996). Now, we are going to introduce two correlation-based quality metrics, namely, correlation quality (CQ) and normalized cross-correlation (NC). These two metrics are defined respectively in Eqs. (6.13) and (6.14).

$$CQ = \frac{\sum_{m=1}^{M}\sum_{n=1}^{N} c(m,n)\cdot e(m,n)}{\sum_{m=1}^{M}\sum_{n=1}^{N} c(m,n)} \qquad (6.13)$$

$$NC = \frac{\sum_{m=1}^{M}\sum_{n=1}^{N} c(m,n)\cdot e(m,n)}{\sum_{m=1}^{M}\sum_{n=1}^{N} c(m,n)^2} \qquad (6.14)$$

Beyond these, there are a few other pixel-based metrics available for image quality estimation. Structural content (SC) is one of them and its computational value is obtained from the following formula,

$$SC = \frac{\sum_{m=1}^{M}\sum_{n=1}^{N} c(m,n)^2}{\sum_{m=1}^{M}\sum_{n=1}^{N} e(m,n)^2} \qquad (6.15)$$

A histogram of a digital image is a discrete function of distinct intensity values, defined as the number of pixels in that image within that particular intensity level (Gonzalez and Woods, 2008). The estimation of the histogram similarity (HS) for a gray-scale image and its corresponding watermarked image could be utilized as another quality metric,

$$HS = \sum_{l=0}^{255} |f_C(Z) - f_E(Z)| \qquad (6.16)$$

where, $f_C(Z)$ is the relative histogram function of level Z in image I; the value of Z varies from 0 to 255 for single-plane image.

An enhanced visual transparency of the watermark is achieved if the values of NC and SC, obtained for the original and water-marked image pair, are close to unity. Larger values of CQ (in excess of 100) and HS (in the order of thousands) are significant for a good imperceptible data hiding scheme.

Finally, we can introduce the structural similarity index (SSIM), developed by Z. Wang et al. (2004), as a quality metric, which is correlated with the perception of the human visual system (HVS). Rather than using the conventional pixel-based error summation techniques, the SSIM is intended to calculate the distortion in any visual stimuli based on a combination of three different image factors, i.e. luminance distortion, contrast distortion, and correlation loss. Hence, for the image pair C and E, the SSIM could be mathematically expressed as

$$\text{SSIM} = f_L(C,E) f_C(C,E) f_S(C,E) \qquad (6.17)$$

where,

$$f_L(C,E) = \frac{2\mu_C\mu_E + Q_1}{\mu_C^2 + \mu_E^2 + Q_2} \qquad (6.18)$$

$$f_C(C,E) = \frac{2\sigma_C\sigma_E + Q_1}{\sigma_C^2 + \sigma_E^2 + Q_2} \qquad (6.19)$$

$$f_S(C,E) = \frac{2\sigma_C\sigma_E + Q_1}{\sigma_C^2 + \sigma_E^2 + Q_2} \qquad (6.20)$$

Eq. (6.18) defines the luminance comparison function to estimate the proximity between the mean intensity, μ_C, for image C and μ_E for image E. The highest value of this factor will be 1 only when $\mu_C = \mu_E$. After obtaining the mean intensity for an image signal, signal contrast could be estimated in the course of applying standard deviation. For the image signal C, the signal contrast (σ_C) is approximated as,

$$\sigma_C = \sqrt{\left(\frac{1}{(M-1)(N-1)}\sum\nolimits_{m=1}^{M}\sum\nolimits_{n=1}^{N}\left(c(m,n)-\mu_C\right)^2\right)} \qquad (6.21)$$

The signal contrast for image E can also be computed as σ_E using Eq. (6.21). Thus, the second factor, given in Eq. (6.19) and termed the contrast comparison function, is calculated from the values of σ_C and σ_E. This function is also equal to its maximum value 1 only when $\sigma_C = \sigma_E$. Finally, each of the image functions is normalized by its own standard deviation to calculate the third factor, called the structure comparison function, as given in Eq. (6.20). Basically, it determines the correlation coefficients for images C and E, considering, σ_{CE} as the covariance between these images. Q_1, Q_2, and Q_3 are three positive constants, used to avoid a null denominator. The values of the SSIM index may vary within the range of 0 and 1, where the upper boundary value 1 is obtained for two identical images and the lower limit value 0 points to the absence of any correlation at all between the two images.

6.3.2 Evaluation of Imperceptibility for the Proposed Watermarking Scheme

The visual transparency of the watermark in the watermarked image is the assessment of imperceptibility, which is achieved by differentiating the watermarked image with respect to the original image using the image quality metrics, as defined in the previous section. The metrics values are computed for all the images from the USC-SIPI image database (modified to be 256×256 grayscale images), looking at the original images and their corresponding watermarked images (obtained through the proposed embedding algorithm) as noisy images. The metrics values, shown in Table 6.1, together indicate the imperceptibility achieved by the proposed algorithm.

The significance of these metrics values is discussed along with their descriptions. In Table 6.1, we find that for almost all the images, the metrics values are very close to the ideal values that could be achieved for two identical images. The high average peak signal-to-noise ratio (PSNR) (obtained as 56.20 dB) with first-rate SSIM values for all the images affirms that the data transparency quality offered by this proposed watermarking algorithm is impressive.

TABLE 6.1 Imperceptibility Assessment Table

(A)

Images	SNR	MSE	PSNR	MD	AD	NAD	NMSE	IF
4.1.01	47.48	0.090	57.84	7	0.0278	0.00047	0.000006	0.999982
4.1.02	41.93	0.135	56.43	7	0.0350	0.00105	0.000017	0.999936
4.1.03	54.33	0.075	59.36	7	0.0265	0.00019	0.000001	0.999996
4.1.04	50.83	0.123	57.07	15	0.0320	0.00029	0.000002	0.999992
4.1.05	54.79	0.070	59.10	7	0.0263	0.00019	0.000001	0.999997
4.1.06	54.23	0.080	58.46	15	0.0260	0.00020	0.000001	0.999996
4.1.07	57.71	0.055	58.80	3	0.0233	0.00013	0.000001	0.999998
4.1.08	56.85	0.062	58.47	7	0.0248	0.00015	0.000001	0.999998
4.2.01	51.80	0.088	58.32	12	0.0270	0.00026	0.000002	0.999993
4.2.02	57.10	0.088	58.66	15	0.0273	0.00013	0.000001	0.999998
4.2.03	43.12	0.889	47.08	15	0.0958	0.00074	0.000005	0.999951
4.2.04	52.18	0.107	57.32	15	0.0296	0.00024	0.000002	0.999994
4.2.05	57.71	0.058	59.68	7	0.0237	0.00013	0.000001	0.999998
4.2.06	49.01	0.250	53.67	15	0.0434	0.00035	0.000002	0.999987
4.2.07	51.78	0.115	56.48	14	0.0313	0.00026	0.000002	0.999993
5.1.09	46.81	0.356	52.40	15	0.0546	0.00043	0.000003	0.999979
5.1.10	47.72	0.368	52.23	15	0.0550	0.00039	0.000003	0.999983
5.1.11	56.81	0.080	58.26	15	0.0235	0.00012	0.000001	0.999998

(Continued)

TABLE 6.1 (CONTINUED) Imperceptibility Assessment Table

(A)

Images	SNR	MSE	PSNR	MD	AD	NAD	NMSE	IF
5.1.12	58.06	0.059	60.10	7	0.0235	0.00013	0.000001	0.999998
5.1.13	57.32	0.105	57.90	3	0.0351	0.00016	0.000001	0.999998
5.1.14	50.63	0.110	57.72	14	0.0300	0.00029	0.000002	0.999991
5.2.08	45.70	0.448	51.62	15	0.0545	0.00044	0.000003	0.999973
5.2.09	51.08	0.265	53.90	15	0.0459	0.00025	0.000001	0.999992
5.2.10	45.78	0.416	51.94	15	0.0607	0.00053	0.000004	0.999974
5.3.01	47.73	0.188	54.61	15	0.0388	0.00044	0.000003	0.999983
5.3.02	45.47	0.225	53.98	15	0.0412	0.00050	0.000005	0.999972
7.1.01	46.09	0.299	50.05	15	0.0487	0.00045	0.000004	0.999975
7.1.02	57.27	0.058	58.69	7	0.0241	0.00014	0.000001	0.999998
7.1.03	53.68	0.078	57.01	7	0.0266	0.00020	0.000001	0.999996
7.1.04	48.56	0.204	53.43	15	0.0396	0.00034	0.000003	0.999986
7.1.05	49.49	0.140	53.88	15	0.0356	0.00034	0.000003	0.999989
7.1.06	48.85	0.120	53.91	7	0.0320	0.00035	0.000003	0.999987
7.1.07	51.77	0.081	55.50	7	0.0279	0.00026	0.000002	0.999993
7.1.08	52.41	0.096	56.02	7	0.0300	0.00024	0.000002	0.999994

(Continued)

TABLE 6.1 (CONTINUED)   Imperceptibility Assessment Table

(A)

Images	SNR	MSE	PSNR	MD	AD	NAD	NMSE	IF
7.1.09	53.11	0.083	57.00	7	0.0273	0.00022	0.000002	0.999995
7.1.10	48.96	0.189	52.38	15	0.0406	0.00034	0.000003	0.999987
7.2.01	42.69	0.090	55.59	15	0.0269	0.00083	0.000016	0.999946
boat.512	55.12	0.058	60.21	7	0.0235	0.00018	0.000001	0.999997
elaine.512	48.25	0.309	52.81	15	0.0479	0.00035	0.000002	0.999985
house	52.48	0.160	55.53	15	0.0368	0.00023	0.000001	0.999994
gray21.512	56.17	0.053	60.86	12	0.0185	0.00015	0.000001	0.999998
numbers.512	51.00	0.108	57.81	14	0.0294	0.00028	0.000002	0.999992
ruler.512	57.37	0.097	58.28	15	0.0281	0.00012	0.000001	0.999998
testpat.1k	57.49	0.035	62.67	3	0.0117	0.00009	0.000001	0.999998

(B)

Images	LMSE	NQM	WDR	CQ	NC	SC	HS	SSIM
4.1.01	0.001086	38.77	47.48	85.63	0.999791	1.000400	1821	0.999966
4.1.02	0.001940	35.92	41.93	63.32	0.999648	1.000641	2297	0.999971
4.1.03	0.000000	42.19	54.33	146.00	0.999838	1.000321	1739	0.999996
4.1.04	0.000360	40.15	50.83	133.32	0.999797	1.000397	2098	0.999983
4.1.05	0.000423	41.91	54.79	153.50	0.999877	1.000243	1723	0.999962
4.1.06	0.000001	41.89	54.23	164.39	0.999856	1.000284	1705	0.999969

(Continued)

TABLE 6.1 (CONTINUED)   Imperceptibility Assessment Table

(B)

Images	SNR	MSE	PSNR	MD	AD	NAD	NMSE	IF
4.1.07	0.000000	43.68	57.71	184.17	0.999934	1.000130	1524	0.999955
4.1.08	0.000000	43.25	56.85	178.25	0.999927	1.000145	1625	0.999957
4.2.01	0.001178	40.58	51.80	128.64	0.999933	1.000128	1768	0.999973
4.2.02	0.000004	43.73	57.10	214.93	0.999939	1.000121	1790	0.999962
4.2.03	0.002608	36.85	43.12	140.78	0.999689	1.000574	6278	0.999960
4.2.04	0.000022	41.18	52.18	142.17	0.999869	1.000255	1939	0.999957
4.2.05	0.000001	43.99	57.71	190.80	0.999938	1.000122	1556	0.999977
4.2.06	0.002323	39.75	49.01	158.76	0.999844	1.000300	2847	0.999974
4.2.07	0.000305	41.10	51.78	143.89	0.999882	1.000230	2049	0.999983
5.1.09	0.000040	39.60	46.81	133.76	0.999815	1.000350	3578	0.999930
5.1.10	0.000047	39.81	47.72	155.11	0.999813	1.000357	3603	0.999972
5.1.11	0.000484	47.35	56.81	199.20	0.999967	1.000064	1541	0.999969
5.1.12	0.000000	44.34	58.06	203.59	0.999942	1.000114	1539	0.999959
5.1.13	0.000000	40.45	57.32	251.28	0.999842	1.000314	2302	0.999916
5.1.14	0.000013	40.27	50.63	121.62	0.999820	1.000352	1963	0.999990
5.2.08	0.006502	40.46	45.70	134.97	0.999727	1.000519	3572	0.999993
5.2.09	0.000954	41.36	51.08	188.20	0.999890	1.000212	3009	0.999983

(Continued)

TABLE 6.1 (CONTINUED)   Imperceptibility Assessment Table

(B)

Images	SNR	MSE	PSNR	MD	AD	NAD	NMSE	IF
5.2.10	0.000973	38.20	45.78	138.44	0.999739	1.000497	3975	0.999973
5.3.01	0.001112	40.22	47.73	124.97	0.999852	1.000280	2541	0.999986
5.3.02	0.003347	38.52	45.47	95.27	0.999788	1.000396	2703	0.999984
7.1.01	0.002545	38.22	46.09	113.56	0.999796	1.000383	3191	0.999951
7.1.02	0.000000	43.85	57.27	178.00	0.999937	1.000125	1581	0.999961
7.1.03	0.000000	41.94	53.68	137.59	0.999889	1.000218	1742	0.999968
7.1.04	0.000836	40.19	48.56	126.25	0.999871	1.000245	2595	0.999961
7.1.05	0.000189	39.86	49.49	117.10	0.999822	1.000344	2336	0.999980
7.1.06	0.000839	39.58	48.85	101.88	0.999856	1.000274	2096	0.999981
7.1.07	0.000072	41.14	51.77	113.04	0.999879	1.000235	1826	0.999985
7.1.08	0.000018	41.31	52.41	131.51	0.999880	1.000234	1968	0.999973
7.1.09	0.000000	41.81	53.11	135.36	0.999902	1.000192	1792	0.999982
7.1.10	0.000237	40.25	48.96	125.07	0.999850	1.000288	2662	0.999955
7.2.01	0.001179	36.42	42.69	51.62	0.999615	1.000716	1762	0.999958
boat.512	0.000000	42.59	55.12	145.78	0.999907	1.000183	1541	0.999971
elaine.512	0.000669	39.32	48.25	151.61	0.999824	1.000338	3139	0.999959
house	0.000010	41.11	52.48	174.99	0.999849	1.000295	2413	0.999938

(*Continued*)

TABLE 6.1 (CONTINUED) Imperceptibility Assessment Table

(B)

Images	SNR	MSE	PSNR	MD	AD	NAD	NMSE	IF
gray21.512	0.000051	46.28	56.17	173.52	0.999999	0.999999	1212	0.999960
numbers.512	0.000164	40.69	51.00	131.13	0.999876	1.000241	1926	0.999996
ruler.512	0.000058	42.42	57.37	234.86	0.999899	1.000200	1840	0.999997
testpat.1k	0.000000	40.27	57.49	158.35	1.000076	0.999846	768	0.999999

6.4 OUTPUTS OF THE WATERMARK EXTRACTING SYSTEM AND ITS EFFICACY IN TERMS OF ROBUSTNESS

As discussed above, robustness is another paramount property of a good watermarking scheme. It is defined as the property of an image to sustain its individuality in a noisy environment. A robust watermarking scheme should attempt to implant the watermark into the cover in such a way that the embedded watermark will be less affected by signal processing attacks (as discussed in Section 1.2.6). Both the embedding process and the extracting mechanism have a distinct contribution in improving the robustness of a watermarking system. An effective extraction logic can retrieve the watermark from a watermarked image with great accuracy. To demonstrate the efficacy of our proposed extraction algorithm, several malicious signal processing attacks are intentionally applied on the watermarked images (obtained through the relative embedding logic) and the extraction operation is performed to retrieve watermark from the noisy images. The extracted watermarks, shown in Figure 6.2, demonstrate the capability of this proposed extraction methodology.

The resemblance of the extracted or recovered watermark to the original mark gives an estimate of the robustness of the system. For this purpose, several quality metrics have been used to determine the dissimilarities in the extracted watermarks compared to the original watermark. Considering the original watermark as W (of size $X \times Y$) and the extracted watermark as W_R, the quality metrics are formed. For a binary watermark, a calculation of the bit error rate (BER) could be very useful to quantify the distortions appearing in the extracted watermark. BER is defined as,

$$\text{BER} = \frac{\text{Total number of erroneous bits}}{\text{Total number of bits in image signal}} \times 100\% \quad (6.22)$$

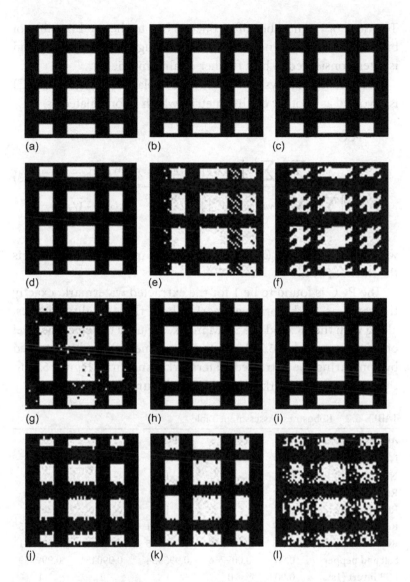

FIGURE 6.2 Extracted watermarks from the received images after applying several attacks: (a) No attack; (b) negative; (c) 90° rotation; (d) 180° rotation; (e) 45° rotation; (f) 10° rotation; (g) salt and pepper; (h) least significant bit (LSB) inversion (1st LSB ↔ 2nd LSB); (i) cropping; (j) erode; (k) dilate; (l) median filtering.

The ratio of the number of erroneous bits to the total number of bits is also known as normalized Hamming distance (NHD) and is often considered to be a further quality metric.

In robustness analysis, the Pearson correlation coefficient (PCC) is involved as another quality metric, which is formulated as,

$$
PCC = \frac{\sum_{x=1}^{X}\sum_{y=1}^{Y}\left[(w(x,y)-\overline{W})(w_R(x,y)-\overline{W_R})\right]}{\sqrt{\left[\sum_{x=1}^{X}\sum_{y=1}^{Y}(w(x,y)-\overline{W})\right]^2\left[\sum_{x=1}^{X}\sum_{y=1}^{Y}(w_R(x,y)-\overline{W_R})\right]^2}}
\tag{6.23}
$$

where, $w(x,y) \in W$, $w_R(x,y) \in W_R$, $\overline{W}$ is the mean of W and $\overline{W_R}$ is the mean of W_R.

The PCC is found to be 1 for the extracted watermark, exactly the same as to the original mark. These three quality metrics, along with the NC and the SSIM index, are used to detect the level of robustness. Table 6.2 shows the experimental results obtained by comparing the extracted watermarks, shown in Figure 6.2, to the original watermark, also shown in Figure 6.2.

TABLE 6.2   Robustness Assessment Table

Attacks	BER	NHD	PCC	NC	SSIM
No attack	0.00	0	1	1	1
Negative	0.00	0	1	1	1
Rotation 90	0.00	0	1	1	1
Rotation 180	0.00	0	1	1	1
Rotation 45	24.37	0.243652	0.893851	0.59375	0.981985
Rotation 10	13.26	0.132568	0.937325	0.764757	0.993481
Salt and pepper	0.98	0.009766	0.995454	0.990451	0.999739
LSB invert (1st LSB↔2nd LSB)	0.00	0	1	1	1
Cropping	0.00	0	1	1	1
Erode	5.49	0.054932	0.965711	0.836806	0.998346
Dilate	5.15	0.051514	0.977801	0.973958	0.998399
Median filtering	11.72	0.117188	0.934359	0.743924	0.995787

Figure 6.2 and Table 6.2 together bear out that the efficacy of this data extracting system is sufficient for an analogous embedding logic, offering enhanced payload with high data transparency.

6.5 COMPARATIVE STUDY ON THE PROFICIENCY OF THE PROPOSED WATERMARKING SYSTEM

A comparative study between the proposed algorithm and some other existing methods is shown in Table 6.3. It is quite normal that the watermark and cover images, used in different proposals, are not identical in size or type. An average value taken from the watermark and the document scale are approximated to give a comparison. Here, it is observed that this projected watermarking

TABLE 6.3 Comparative Study of Some Existing Image Watermarking Systems

Sl. No.	Method	PSNR (dB)	Payload Capacity (bpp)
1.	Proposed method	56.20	2.30
2.	Adaptive LSB replacement (Sinha Roy et al., 2018)	53.5	2.17
3.	Magic cube-based information hiding (Wu et al., 2016)	45.15	2
4.	LSB replacement method (Goyal and Kumar, 2014)	54.8	1
5.	Reversible data hiding scheme (Gui et al., 2014)	34.26	1
6.	Matrix encoding-based watermarking (Verma and Yadav, 2013)	55.91	1.99
7.	Salient region watermarking (Wong et al., 2013)	45.83	0.58
8.	Adaptive pixel pair-matching scheme (Hong and Chen, 2012)	40.97	3
9.	Discrete wavelet transform (DWT)- and singular value decomposition (SVD)-based watermarking (Majumdar et al., 2011)	41.72	0.375
10.	Pair-wise LSB-matching (Xu et al., 2010)	35.05	2.25
11.	Internet protocol (IP) LSB (Yang, 2008)	35.07	4
12.	Optimal LSB pixel adjustment (Yang, 2008)	34.84	4
13.	Mielikainen's method (Mielikainen, 2006)	33.05	2.2504
14.	Pair-wise logical computation (PWLC) data hiding (Tsai et al., 2005)	48.35	0.68
15.	Exhaustive LSB substitution (Chang et al., 2003)	38.34	3

technique gives a much better PSNR, with a remarkable data hiding capacity, than most of the schemes. Only a few techniques can provide improved payloads, but these are lagging behind the proposed method in terms of the PSNR. Moreover, the previous section has confirmed that the robustness of the proposed framework is also noteworthy. Thus, from the overall system evaluation taken together with these comparison results, we can conclude that the proposed watermarking algorithm is optimized in terms the three major properties (i.e. data hiding capacity, imperceptibility, and robustness) required for any digital image watermarking scheme.

6.6 AN EVALUATION OF HARDWARE SYSTEMS FOR THE PROPOSED WATERMARKING LOGIC

The effectiveness of any hardware system is evaluated in terms of its accuracy in performance, execution time, required space, and computational cost. Here, the FPGA-based hardware implementations of the proposed embedding and extracting systems were executed using Xilinx ISE 13.2. The Spartan Series of FPGA was used in regulating the compactness of the design and xc6slx4-3tqg144 was chosen as the target device. As a hardware description language (HDL), VHDL is assigned, where 'V' indicates the first letter of the acronym 'VHSIC, which stands for very high speed integrated circuit (Bhasker, 2001). The behavioral simulation was accomplished using the Xilinx ISim for the synthesizability; and, after execution, the top-level RTL (register-transfer level) schematics of the watermark embedding and extracting systems were obtained, as shown in Figures 6.3 and 6.4 respectively.

To verify the FPGA model of the embedding system, a simulation process was performed using a 16-bit sequence constructed from the watermark provided in Figure 4.8b as the binary watermark. The other inputs received several 8-bit image pixels as grayscale cover image pixels together with the analogous HCM pixels selected from the cover image given in Figure 4.8a and its relative HCM shown in Figure 4.8d. These input vectors are

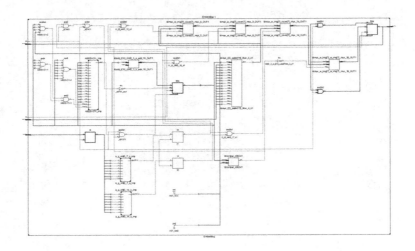

FIGURE 6.3   RTL schematic diagram for the watermark embedding system.

supplied through test benches written in VHDL. The selection of the cover pixel values and their corresponding HCM pixel values is consciously carried out in such a way that the simulation results can be observed for all conditions of the proposed data implanting logic. The simulation results obtained for watermark inserting process are shown in Figure 6.5. Here we can see how the watermark bits are adaptively embedded in the cover image pixels to produce the watermarked pixels. The watermarked pixels, acquired in the bit-inserting process, are consecutively applied to the input of the extraction system. Based on the HCM values, parallel to the provided watermarked pixels, the watermark retrieving block extracts the watermark bits in successive instants of time. The bit sequence obtained with valid '1' is exactly identical to the bit stream used as the watermark in the embedding process. The simulation output for the watermark extracting system is presented in Figure 6.6.

With a 50 ns time period of the clock pulse, 1500 ns is the required time to execute each of the embedding and extracting systems, considering the input image dimensions and all other

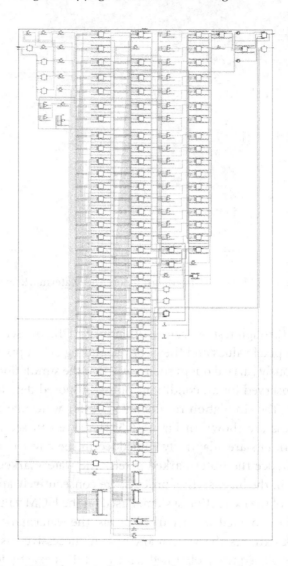

FIGURE 6.4   RTL schematic diagram for the watermark extracting system.

FIGURE 6.5   Behavioral simulation results for the watermark embedding system.

FIGURE 6.6   Behavioral simulation results for the watermark extracting system.

TABLE 6.4   Device Utilization Summary for (A) the Embedding System and (B) the Extracting System, Executed through Xilinx ISE 13.2

Logic Components	Available	Used	Utilization (%)
(A) Embedding system			
Number of slice registers	4800	38	0
Number of slice LUTs	2400	45	1
Number of fully used LUT-flip-flop (LUT-FF) pairs	56	28	50
Number of bonded Input Output Blocks (IOBs)	102	40	39
Number of global clock buffers (BUFG)/ global clock buffer controls (BUFGCTRL)	16	2	12
(B) Extracting system			
Number of slice registers	4800	38	0
Number of slice LUTs	2400	45	1
Number of fully used LUT-FF pairs	55	28	50
Number of bonded IOBs	102	40	39
Number of BUFG/BUFGCTRLs	16	2	12

software specifications remain the same as those described. From the device utilization summary table (Table 6.4), it is clear that a small percentage of the available techniques have been used to implement both the insertion and extraction processes. This signifies that the execution time and cost are both reduced for these hardware systems. Thus, on-chip implementation can be achieved with low design complexity, and the space requirements will also be very low.

CHAPTER 7

Conclusion

COPYRIGHT PROTECTION FOR MULTIMEDIA objects is an emerging issue in digital signal processing and in present-day data communications. Digital watermarking has been introduced as a perfect solution for this issue. For any digital watermarking scheme, imperceptibility, robustness, and data hiding capacity are the three vital properties, however, by their very nature, they are in conflict with each other. To overcome this, heuristic watermarking algorithms have been developed for use in either the spatial or the frequency domain. Today, the practice of digital watermarking has become a recognized domain for research and development. This book has discussed the emergence of digital watermarking as a copyright protection tool, its basic working principles, applications, and challenges, and finally has focused on the development of an image watermarking scheme. A spatial domain-based novel digital image watermarking technique has been proposed here as a copyright protection tool, using some intelligent techniques, e.g. saliency detection and data clustering, to enhance data transparency according to the human visual system. This proposed technique embeds the watermark into the cover image pixels through an adaptive least significant bit (LSB) replacement technique. Bit replacement is performed in such a manner that the maximum amount of information can be stored in the busiest regions. A spectral residual-based saliency map algorithm is involved in saliency detection and a K-means clustering algorithm is used to group the image pixels, based on the level of saliency. As a consequence, high data transparency is achieved with improved payload. The effectiveness of this proposed method is approximated in terms of imperceptibility and robustness, using several image quality metrics. This method is also compared with some other existing systems, considering data transparency and hiding capacity. Lastly, the hardware realization of the proposed embedding and extracting systems is performed through FPGA-design modes. From the comparative study and analysis of the system outputs, it is concluded that this approach offers several advantages. These can be summarized as follows:

- Utilizing the imperfect nature of the human visual system (HVS) in data implantation, this algorithm delivers better visual transparency.

- An improved robustness is obtained against several signal processing and geometrical attacks.

- The hiding capacity is also very good.

- This proposed system is more flexible in nature, as the payload can be varied up to a certain range.

- The involvement of automated data clustering and the generation of the hiding capacity map have become generic in nature using this system. Varying the number of clusters, data hiding could be modified according to the user's requirements.

- Finally, being developed in the spatial domain, the computation cost and complexity are optimized in designing the hardware architecture for this proposed watermark embedding and extracting system. Moreover, the runtime, required for the hardware systems, is also very good.

Therefore, the watermarking methodology developed in this book can offer guidance in developing new watermarking techniques as well as in improving the existing models. Apart from the proposed methodology, this book also introduces many important factors related to image watermarking concepts, indicating the way forward for further research and development.

For spatial watermarking schemes, the data hiding capacity can be increased up to a specific level as a maximum. Otherwise, the imperceptibility could be hampered. We could consider the development of other data hiding methods in order to increase payload capacity with the same or better data transparency. Moreover, in the spatial domain, robustness is generally poor with respect to that found using frequency domain techniques. Therefore, the intelligent techniques favored could be combined with certain frequency domain approaches, so that better image quality might be achieved together with improved robustness.

The data embedding methodologies can also be improved in the spatial domain. For instance, the utilization of reversible logic (Dueck and Maslov, 2003; Maslov et al., 2005; Miller et al., 2003; Wille et al., 2008) could offer a better way to implant watermarks, either through spatial domain- or through transform domain-based practices, providing a balance between data transparency and payload capacity. Similarly, we can test some other saliency map generation algorithms and clustering models in order to modify the contributions of these intelligent techniques. In other words, this watermarking technique can be improved with the melioration of certain HVS models.

Currently, a binary watermark image is typically embedded into a grayscale image. Using this basic concept, the watermark insertion procedure for color images could also be executed. Instead of images, other multimedia objects, such as text, sound, video, etc. could even be considered as the host or the cover. The nature of the watermark objects themselves can also be varied in further observations.

Hardware implementation is performed using FPGA as this offers a low-cost solution, however, it can also be executed using application-specific integrated circuit (ASIC) functions, although the computational cost and device utilization are much less for this proposed FPGA architecture. Quantum-based hardware implementations (Sinha Roy et al., 2018; Lukac et al., 2003) should be analyzed in further developments, as the processing speed is very high and power dissipation is very low for these types of designs.

We have introduced numerous image quality metrics to evaluate the system performance. As well as considering these quality metrics individually, it could be effective to produce a generic testing architecture. In addition, the utilization of some optimization techniques, such as genetic algorithms (Grefenstette, 1986; Srinivas and Deb, 1994), differential evolution algorithms (Storn and Price, 1997), etc. could be used for designing a watermarking system in order to obtain the optimum results according to different requirements or purposes. These concepts also suggest the formulation of a universal testing parameter, so that the proficiency of any watermarking schemes can be assessed using a standardized scale. For example, a proper benchmark, explored through ITU-R Rec. 500 quality rating (Kutter and Petitcolas, 1999) on a scale from 1 to 5, is usually considered a well-known quantitative parameter to justify the visual transparency of the embedded watermark for any data hiding scheme.

This proposed watermarking system has also undergone testing using USC-SIPI database images and was rated as 4. This rating again demonstrates that hardly any visual distortion could be recognized in an aesthetic sense for this proposed watermarking scheme.

In conclusion, this book has illustrated the development of a novel intelligent technique-based image watermarking scheme as a copyright protection tool for digital images, and, parallel to that, it has unveiled an important emerging field for further research and development.

References

Al-Nabhani, Y., Jalab, H. A., Wahid, A., and Noor, R., "Robust Watermarking Algorithm for Digital Images Using Discrete Wavelet and Probabilistic Neural Network," *Journal of King Saud University – Computer and Information Sciences 27*, no. 4 (2015): 393–401.

Anderson, R. J., "Liability and Computer Security: Nine Principles," in *Computer Security – ESORICS 94: Third European Symposium on Research in Computer Security*, LNCS, Vol. 875, Springer, Berlin, Germany (1994): 231–245.

Anderson, R. J. and Petitcolas, F. A. P., "On the Limits of Steganography," *IEEE Journal on Selected Areas in Communications 16*, no. 4 (1998): 474–481.

Anitha, J. and Immanuel Alex Pandian, S., "A Color Image Digital Watermarking Scheme Based on SOFM," *IJCSI International Journals of Computer Science Issues 7*, no. 5 (2010): 302–309.

Anumol, T. J., Binson, V. A., and Rasheed, S., "FPGA Implementation of Low Power, High Speed, Area Efficient Invisible Image Watermarking Algorithm for Images," *International Journal of Structural Engineering 4*, no. 8 (2013): 1–6.

Arabie, P. and Hubert, L., "Cluster Analysis in Marketing Research," in Richard P. Bagozzi (Ed.), *Advanced Methods in Marketing Research*, Blackwell, Oxford, UK (1994): 160–189.

Baldi, P. and Hatfield, G., *DNA Microarrays and Gene Expression*, Cambridge University Press, Cambridge, UK (2002).

Banerjee, S., Chakraborty, S., Dey, N., Kumar Pal, A., and Ray, R., "High Payload Watermarking Using Residue Number System," *International Journal of Image, Graphics and Signal Processing 7*, no. 3 (2015): 1–8.

Barr, T., *Invitation to Cryptography*, Prentice Hall, Upper Saddle River, NJ (2002).

Basu, A. and Sarkar, S. K., "On the Implementation of Robust Copyright Protection Scheme Using Visual Attention Model," *Information Security Journal: A Global Perspective* 22, no. 1 (2013): 10–20.

Basu, A., Das, T. S., and Sarkar, S. K., "On the Implementation of an Information Hiding Design Based on Saliency Map," in *International Conference on Image Information Processing*, IEEE, Piscataway, NJ (2011a): 1–6.

Basu, A., Das, T. S., and Sarkar, S. K., "Robust Visual Information Hiding Framework Based on HVS Pixel Adaptive LSB Replacement (HPALR) Technique," *International Journal of Imaging and Robotics* 6 (2011b): 71–98.

Basu, A., Sinha Roy, S., and Chattopadhayay, A., "Implementation of a Spatial Domain Salient Region Based Digital Image Watermarking Scheme," in *International Conference Research in Computational Intelligence and Communication Networks (ICRCICN)*, IEEE, Piscataway, NJ (2016): 269–272.

Basu, A., Sinha Roy, S., and Sarkar, S., "FPGA Implementation of Saliency Based Watermarking Framework," in *6th International Conference on Computers and Devices for Communication (CODEC)* (2015). Available: https://drive.google.com/file/d/1DN3 jhKEWKnDdp_WYDI67mYB8nlm7eaLy/view.

Berghel, H., "Watermarking Cyberspace," *Communications of the ACM* 40, no. 11 (1997): 19–24.

Bezdeck, J. C., Ehrlich, R., and Full, W., "FCM: Fuzzy C-Means Algorithm," *Computer and Geoscience* 10, no. 2–3 (1984): 191–203.

Bhasker, J., "A VHDL Synthesis Primer," 2E, BS Publication, Andhra Pradesh, India (2001).

Bhattacharya, A., Palit, S., Chatterjee, N., and Roy, G., "Blind Assessment of Image Quality Employing Fragile Watermarking," in *7th International Symposium on Image and Signal Processing and Analysis (ISPA)*, IEEE, Piscataway, NJ (2011): 431– 436.

Bhattacharya, S., "Survey on Digital Watermarking – A Digital Forensics and Security Application," *International Journal of Advanced Research in Computer Science and Software Engineering* 4, no. 11 (2014): 1–7.

Borda, M., *Fundamentalism Information Theory and Coding*, Springer Science and Business Media, Chennai, India (2011): 178–180.

Borji, A., "Boosting Bottom-Up and Top-Down Visual Features for Saliency Estimation," in *Proceedings of the IEEE Computer Vision and Pattern Recognition (CVPR)*, IEEE, Piscataway, NJ (2012): 438–445.

Carmi, R. and Itti, L., "Visual Causes versus Correlates of Attentional Selection in Dynamic Scenes," *Vision Research* 46, no. 26 (2006): 4333–4345.

Chan, H.-T., Hwang, W., and Cheng, C., "Digital Hologram Authentication Using a Hadamard– Based Reversible Fragile Watermarking Algorithm," *Journal of Display Technology* 11, no. 2 (2015): 193–203.

Chang, C. C., Hsiao, J. Y., and Chan, C. S., "Finding Optimal Least-Significant-Bit Substitution in Image Hiding by Dynamic Programming Strategy," *Pattern Recognition* 36, no. 7 (2003): 1583–1595.

Chen, D.-Y., Ouhyoung, M., and Wu, J.-L., "A Shift-Resisting Public Watermark System for Protecting Image Processing Software," *IEEE Transactions on Consumer Electronics* 46, no. 3 (2000): 404–414.

Chen, W. C. and Wang, M. S., "A Fuzzy c-Means Clustering Based Fragile Watermarking Scheme for Image Authentication," *Expert Systems with Applications* 36, no. 2 (2009): 1300–1307.

Chou, C. M. and Tseng, D. C., "A Public Fragile Watermarking Scheme for 3D Model Authentication," *Computer-Aided Design* 38, no. 11 (2006): 1154–1165.

Chua, H. F., Boland, J. E., and Nisbett, R. E., "Cultural Variation in Eye Movements during Scene Perception," *Proceedings of the National Academy of Sciences of the United States of America* 102, no. 35 (2005): 12629–12633.

Colonna, F., *Hypnerotomachia Poliphili: The Dream of Polia's Lover*, 1E, Aldus Manutius (1999).

Cox, I. J., Kilian, J., Leighton, F. T., and Shamoon, T., "Secure Spread Spectrum Watermarking for Images, Audio and Video," in *Proceedings of 3rd IEEE International Conference on Image Processing*, Vol. 3, IEEE, Piscataway, NJ (1996): 243–246.

Cox, I. J., Kilian, J., Leighton, F. T., and Shamoon, T., "Secure Spread Spectrum Watermarking for Multimedia," *IEEE Transactions on Image Processing* 6, no. 12 (1997): 1673–1687.

Dueck, G. W. and Maslov, D., "Reversible Function Synthesis with Minimum Garbage Outputs," in *Proceedings of the 6th International Symposium on Representations and Methodology of Future Computing Technologies*, Hasso Plattner Institut, Trier, Germany (2003): 154–161.

El-Araby, W. S., Madian, A. H., Ashour, M. A., and Wahdan, A. M., "Hardware Realization of DC Embedding Video Watermarking Technique Based on FPGA," in *International Conference on Microelectronics (ICM)*, IEEE, Piscataway, NJ (2010): 463–466.

Fan, Y. C., Chiang, A., and Shen, J. H., "ROI-Based Watermarking Scheme for JPEG 2000," *Circuits, Systems and Signal Processing* 27, no. 5 (2008): 763–774.

Felzenszwalb, P. F., Girshick, R. B., McAllester, D., and Ramanan, D., "Object Detection with Discriminatively Trained Part Based Models," *IEEE Transactions on Pattern Analysis and Machine Intelligence* 32, no. 9 (2010): 1627–1645.

Field, D. J. and Brady, N., "Visual Sensitivity, Blur and the Sources of Variability in the Amplitude Spectra of Natural Scenes," *Vision Research* 37, no. 23 (1997): 3367–3383.

Friedman, G. L., "The Trustworthy Digital Camera: Restoring Credibility to the Photographic Image," *IEEE Transactions on Consumer Electronics* 39, no. 4 (1993): 905–910.

Frigui, H. and Krishnapuram, R., "A Robust Competitive Clustering Algorithm with Applications in Computer Vision," *IEEE Transactions on Pattern Analysis and Machine Intelligence* 21, no. 5 (1999): 450–465.

Ganguly, A. B., *Fine Arts in Ancient India*, Abhinav Publications, New Delhi, India (1979).

Gerzon, M. A. and Graven, P. G., "A High-Rate Buried-Data Channel for Audio CD," *Journal of the Audio Engineering Society* 43, no. 1–2 (1995): 3–22.

Ghosh, S., Biswas, A., Maity, S. P., and Rahaman, H., "Design of a Low Complexity and Fast Hardware Architecture for Digital Image Watermarking in FWHT Domain on FPGA," in *ISED 2014: Fifth International Symposium on Electronic System Design*, IEEE, Piscataway, NJ (2014): 15–17.

Goferman, S., Zelnik-Manor, L., and Tal, A., "Context-Aware Saliency Detection," *IEEE Transactions on Pattern Analysis and Machine Intelligence* 34, no. 10 (2012): 1915–1926.

Gonzalez, R. C. and Woods, R. E., *Digital Image Processing*, 3E, Pearson, Upper Saddle River, NJ (2008).

Goyal, R. and Kumar, N., "LSB Based Digital Watermarking Technique," *International Journal of Application or Innovation in Engineering and Management* 3, no. 9 (2014): 15–18.

Greenspan, H., Belongie, S., Goodman, R., Perona, P., Rakshit, S., and Anderson, C. H., "Overcomplete Steerable Pyramid Filters and Rotation Invarience," in *Proceedings of the IEEE Conference on Computer Vision and Pattern Recognition (CVPR)*, IEEE, Piscataway, NJ (1994): 222–228.

Grefenstette, J. J., "Optimization of Control Parameters for Genetic Algorithms," *IEEE Transactions on Systems, Man, and Cybernetics* 16, no. 1 (1986): 122–128.

Gui, X., Li, X., and Yang, B., "A High Capacity Reversible Data Hiding Scheme Based on Generalized Prediction-Error Expansion and Adaptive Embedding," *Signal Processing* 98 (2014): 370–380.

Guo, H., Li, Y., Liu, A., and Jajodia, S., "A Fragile Watermarking Scheme for Detecting Malicious Modifications of Database Relations," *Information Sciences* 176, no. 10 (2006): 1350–1378.

Harel, J., Koch, C., and Perona, P., "Graph-Based Visual Saliency," in *Proceedings of the Neural Information Processing Systems*, MIT Press, Cambridge, MA (2006): 545–552.

Hayhurst, J. D. *The Pigeon Post into Paris 1870–1871*, Author, Ashford, UK (1970).

Hernandez Martin, J. R. and Kutter, M., "Information Retrieval in Digital Watermarking," *IEEE Communications Magazine* 39, no. 8 (2001): 110–116.

Herodotus, *The Histories*, J. M. Dent, London, UK (1992).

Ho, A. T. S., Zhu, X., Vrusias, B., and Armstrong, J., "Digital Watermarking and Authentication for Crime Scene Analysis," in *2006 IET Conference on Crime and Security*, Institution of Engineering and Technology, London, UK (2006).

Homer, *The Iliad* (trans. R. Fragels), Penguin, Middlesex, UK (1972).

Hong, W. and Chen, T. S., "A Novel Data Embedding Method Using Adaptive Pixel Pair Matching," *IEEE Transactions on Information Forensics and Security* 7, no. 1 (2012): 176–184.

Hore, A. and Ziou, D., "Image Quality Metrics: PSNR vs. SSIM," in *2010 20th International Conference on Pattern Recognition*, IEEE, Piscataway, NJ (2010): 2366–2369.

Hou, X. and Zhang, L., "Saliency Detection: A Spectral Residual Approach," in *2007 IEEE Conference on Computer Vision and Pattern Recognition*, IEEE, Piscataway, NJ (2007): 1–8.

Hou, X., Harel, J., and Koch, C., "Image Signature: Highlighting Sparse Salient Regions," *IEEE Transactions on Pattern Analysis and Machine Intelligence* 34, no. 1 (2012): 194–201.

Hu, J., Ray, B. K., and Singh, M., "Statistical Methods for Automated Generation of Service Engagement Staffing Plans," *IBM Journal of Research and Development* 51, no. 3 (2007): 281–293.

Huang, J. and Shi, Y. Q., "Adaptive Image Watermarking Scheme Based on Visual Masking," *Electronics Letters* 34, no. 8 (1998): 748–750.

Husain, F., Khan, E., and Farooq, O., "Digital Image Watermarking Using Combined DWT and DFRFT," *International Journal of Computer Applications* 60, no. 11 (2012): 17–25.

Hwai-Tsu, H. and Ling-Yuan, H., "A Mixed Modulation Scheme for Blind Image Watermarking," *International Journal of Electronics and Communications* 70, no. 2 (2016): 172–178.

Itti, L. and Koch, C., "A Saliency-Based Search Mechanism for Overt and Covert Shifts of Visual Attention," *Vision Research* 40, no. 10–12 (2000): 1489–1506.

Itti, L., Koch, C., and Niebur, E., "A Model of Saliency Based Visual Attention for Rapid Scene Analysis," *IEEE Transactions on Pattern Analysis and Machine Intelligence* 20, no. 11 (1998): 1254–1259.

Jain, A. K., "Data Clustering: 50 Years Beyond K-Means," *Pattern Recognition Letters* 31, no. 8 (2010): 651–666.

Jain, A. K. and Flynn, P., "Image Segmentation Using Clustering," in *Advances in Image Understanding*, IEEE Computer Society, Washington, DC (1996): 65–83.

Johnson, N. F., "In Search of the Right Image: Recognition and Tracking of Images in Image Databases, Collections, and the Internet. Technical Report," George Mason University, Center for Secure Information Systems, Fairfax, VA (1999).

Judd, T., Ehinger, K., Durand, F., and Torralba, A., "Learning to Predict Where Humans Look," in *2009 IEEE International Conference on Computer Vision*, IEEE, Piscataway, NJ (2009).

Kahn, D., *Codebreakers: Story of Secret Writing*, Macmillan, New York (1967).

Kahn, D., "The History of Steganography," in *Proceedings of the First International Workshop on Information Hiding, Cambridge, UK*, LNCS, Vol. 1174, Springer, Berlin, Germany (1996): 1–5.

Kang, S.-M. and Leblebici, Y., *CMOS Digital Integrated Circuit*, 3E, Tata McGraw-Hill, New Delhi, India (2003).

Kankanahalli, M. S., Rajmohan, and Ramakrishnan, K. R., "Adaptive Visible Watermarking of Images," in *Proceedings of the IEEE International Conference on Multimedia Computing Systems, (ICMCS-1999)*, Cento Affari, Florence, Italy (1999): 568–573.

Kannammal, A., Pavithra, K., and Subha Rani, S., "Double Watermarking of DICOM Medical Images Using Wavelet Decomposition Technique," *European Journal of Scientific Research* 70, no. 1 (2012): 46–55.

Katzenbeisser, S. and Petitcolas, F. A. P., *Information Hiding Techniques for Steganography and Digital Watermarking*, Artech House, Boston, MA (2000).

Kaur, G. and Kaur, K., "Image Watermarking Using LSB (Least Significant Bit)," *International Journal of Advanced Research in Computer Science and Software Engineering* 3, no. 4 (2013): 358–361.

Kavadia, C. and Shrivastava, V., "A Literature Review on Water Marking Techniques," *International Journal of Scientific Engineering and Technology* 1, no. 4 (2012): 8–11.

Khoshki, R. M., Wang, S., Oweis, S., Pappas, G., and Ganesan, S., "FPGA Hardware Based Implementation of an Image Watermarking System," *International Journal of Advance Research in Computer and Communication Engineering* 3, no. 5 (2014): 6400–6405.

Kim, Y. S., Kwon, O. H., and Park, R. H., "Wavelet Based Watermarking Method for Digital Images Using the Human Visual System," *Electronics Letters* 35, no. 6 (1999): 466–468.

Koch, C. and Ullman, S., "Shifts in Selective Visual Attention: Towards the Underlying Neural Circuitry," *Human Neurobiology* 4, no. 4 (1985): 219–227.

Kootstra, G., Nederveen, A., and Boer, B. D., "Paying Attention to Symmetry," in *British Machine Vision Conference*, BMVA Press, Leeds, UK (2008): 1115–1125.

Kundur, D. and Hatzinakos, D., "Digital Watermarking for Telltale Tamper Proofing and Authentication," *Proceedings of the IEEE* 87, no. 7 (1999): 1167–1180.

Kutter, M., *Digital Watermarking: Hiding Information in Images*, Ph.D thesis 2049, Swiss Federal Institute of Technology, Lausanne, Switzerland (1999).

Kutter, M. and Petitcolas, F. A. P., "A Fair Benchmark for Image Watermarking Systems," in *Electronic Imaging '99, Security and Watermarking of Multimedia Contents*, Vol. 3657, The International Society for Optical Engineering, San Jose, CA (1999): 226–239.

Lande Pankaj, U., Talbar, S. N., and Shinde, G. N., "FPGA Implementation of Image Adaptive Watermarking Using Human Visual Model," *International Conference on Computer Science and Engineering. Journal of Programmable Device, Circuit, and Systems (ICGST-PDCS)* 9, no. 1 (2009): 17–22.

Latif, A., "An Adaptive Digital Image Watermarking Scheme Using Fuzzy Logic and Tabu Search," *The Journal of Information Hiding and Multimedia Signal Processing* 4, no. 4 (2013): 250–271.

Lee, H. K., Kim, H. J., Kwon, S. G., and Lee, J. K., "ROI Medical Image Watermarking Using DWT and Bit-Plane," in *Asia-Pacific Conference on Communications*, IEEE, Piscataway, NJ (2005): 512–515.

Li, J., Dong, C., Han, X.-H., and Chen, Y.-W., "DFT Based Multiple Watermarks for Medical Image Robust to Common and Geometrical Attacks," in *6th International Conference on New Trends in Information Science and Service Science and Data Mining (ISSDM)*, IEEE, Piscataway, NJ (2012a): 472–477.

Li, W., Yang, C., Li, C., and Yang, Q., "JND Model Study in Image Watermarking," in *Advances in Multimedia, Software Engineering and Computing*, Vol. 2 (2012b): 535–543.

Lin, C.-Y. and Chang, S.-F., "Issues for Authenticating MPEG Video," in *Proceedings of the SPIE, Security and Watermarking of Multimedia Contents*, Vol. 3657 (1999): 54–56.

Lin, T. C. and Lin, C. M., "Wavelet Based Copyright Protection Scheme for Digital Images Based on Local Features," *Information Sciences* 179, no. 19 (2009): 3349–3358.

Lukac, M., Perkowski, M., Goi, H., Pivtoraiko, M., Yu, C. H., Chung, K., Jeech, H., Kim, B.-G., and Kim, Y.-D., "Evolutionary Approach to Quantum and Reversible Circuits Synthesis," *Artificial Intelligence Review* 20, no. 3/4 (2003): 361–417.

Maity, H. K. and Maity, S. P., "FPGA Implementation of Reversible Watermarking in Digital Images Using Reversible Contrast Mapping," *Journal of Systems and Software* 96 (2014): 93–104.

Majumdar, S., Das, T. S., and Sarkar, S. K., "DWT and SVD Based Image Watermarking Scheme Using Noise Visibility and Contrast Sensitivity," in *International Conference on Recent Trends in Information Technology*, IEEE, Piscataway, NJ (2011): 938–942.

Mano, M. M. and Ciletti, M. D., *Digital Design*, 5E, Pearson, Upper Saddle River, NJ (2013).

Mao, W., *Modern Cryptography*, Prentice Hall, Upper Saddle River, NJ (2004).

Maslov, D., Dueck, G. W., and Miller, D. M., "Synthesis of Fredkin-Toffoli Reversible Networks," *IEEE Transactions on Very Large Scale Integration* 13, no. 6 (2005): 765–769.

Merriam-Webster Online Dictionary, *Cluster Analysis* (2008). Available: https://www.merriam-webster.com/.

Mielikainen, J., "LSB Matching Revisited," *IEEE Signal Processing Letters* 13, no. 5 (2006): 285–287.

Miller, D. M., Maslov, D., and Dueck, G. W., "A Transformation Based Algorithm for Reversible Logic Synthesis," in *Proceedings 2003. Design Automation Conference*, IEEE, Piscataway, NJ (2003): 318–323.

Mitchell, T. *Machine Learning*, McGraw-Hill, Boston, MA (1997).

Mohammed, A. A. and Sidqi, H. M., "Robust Image Watermarking Scheme Based on Wavelet Technique," *International Journal of Computer Science and Security* 5, no. 4 (2011): 394–404.

Mohanty, S. P., "Digital Watermarking: A Tutorial Review" (1999). Available: http://informatika.stei.itb.ac.id/~rinaldi.munir/Kripto grafi/WMSurvey1999Mohanty.pdf.

Mohanty, S. P., "A Secure Digital Camera Architecture for Integrated Real-Time Digital Rights Management," *Journal of Systems Architecture* 55, no. 10–12 (2009): 468–480.

Mohanty, S. P. and Bhargava, B. K., "Invisible Watermarking Based on Creation and Robust Insertion-Extraction of Image Adaptive Watermarks," *ACM Transactions on Multimedia Computing, Communications, and Applications* 5, no. 2 (2008): 1–22.

Mohanty, S. P., Kumara C. R., and Nayak, S., "FPGA Based Implementation of an Invisible-Robust Image Watermarking Encoder," in *CIT'04 Proceedings of the 7th International Conference on Intelligent Information Technology*, Springer, Berlin, Germany (2004): 344–353.

Mohanty, S. P., Parthasarathy, G., Elias, K., and Nishikanta, P. A., "Novel Invisible Color Image Watermarking Scheme Using Image Adaptive Watermark Creation and Robust Insertion-Extraction," in *Proceedings of the 8th IEEE International Symposium on Multimedia (ISM)*, IEEE Computer Society, Los Alamitos, AZ (2006): 153–160.

Mohanty, S. P., Ramakrishnan, K., and Kankanhalli, M., "A Dual Watermarking Technique for Images," in *Proceedings of the 7th ACM Integracija Multimedia Conference*, ACM Multimedia, Orlando, FL (1999): 49–51.

Nayak, M. R., Bag, J., Sarkar, S., and Sarkar, S. K., "Hardware Implementation of a Novel Watermarking Algorithm Based on Phase Congruency and Singular Value Decomposition Technique," *AEU – International Journal of Electronics and Communication*, Vol. 71 (2017): 1–8.

Ni, R. and Ruan, Q., "Region of Interest Watermarking Based on Fractal Dimension," in *Proceedings of the 18th International Conference on Pattern Recognition*, Vol. 3, IEEE, Piscataway, NJ (2006): 934–937.

Nikolaidis, N. and Pitas, I., "Copyright Protection of Images Using Robust Digital Signatures," in *1996 IEEE International Conference on Acoustics, Speech, and Signal Processing. Conference Proceedings*, Vol. 4, IEEE, Piscataway, NJ (1996): 2168–2171.

Nikolaidis, A. and Pitas, I., "Region-Based Image Watermarking," *IEEE Transactions on Image Processing: A Publication of the IEEE Signal Processing Society* 10, no. 11 (2001): 1726–1740.

Nikolaidis, N. and Pitas, I., "Robust Image Watermarking in Spatial Domain," *Signal Processing* 66, no. 3 (1998): 385–403.

Niu, Y., Kyan, M., Ma, L., Beghdadi, A., and Krishnan, S., "A Visual Saliency Modulated Just Noticeable Distortion Profile for Image Watermarking," in *19th European Signal Processing Conference (EUSIPICO 2011)*, EURASIP, Barcelona, Spain (2011): 2039–2043.

Patra, J. C., Phua, J. E., and Rajan, D., "DCT Domain Watermarking Scheme Using Chinese Remainder Theorem for Image Authentication," in *IEEE International Conference on Multimedia and Expo*, EAAA, Piscataway, NJ (2010): 111–116.

Petitcolas, F. A. P., Anderson, R. J., and Kuhn, M. G., "Information Hiding – A Survey," *Proceedings of the IEEE* 87, no. 7 (1999): 1062–1078.

Pfitzmann, B., "Information Hiding Terminology," in *Proceedings of the First International Workshop on Information Hiding*, LNCS, Vol. 1174, Springer, Berlin, Germany (1996): 347–350.

Pitas, I., "A Method for Signature Casting on Digital Images," in *Proceedings of the IEEE International Conference on Image Processing*, Vol. 3, EAAA, Piscataway, NJ (1996): 215–218.

Qianli, Y. and Yanhong, C., "A Digital Image Watermarking Algorithm Based on Discrete Wavelet Transform and Discrete Cosine Tranform," in *2012 International Symposium on Information Technology in Medicine and Education*, IEEE, Piscataway, NJ (2012): 1102–1105.

Que, Z., Zhu, Y., Wang, X., Yu, J., Huang, T., Zheng, Z., Yang, L., Zhao, F., and Fu, Y., "Implementing Medical CT Algorithms on Stand-Alone FPGA Based Systems Using an Efficient Workflow with Xilinx System Generator and Simulink," in *2010 10th IEEE International Conference on Computer and Information Technology*, IEEE, Piscataway, NJ (2010): 2391–2396.

Ram, B., "Digital Image Watermarking Technique Using Discrete Wavelet Transform and Discrete Cosine Transform," *International Journal of Advancements in Research and Technology* 2, no. 4 (2013): 19–27.

Razafindradina, H. B. and Karim, A. Md., "Blind and Robust Images Watermarking Based on Wavelet and Edge Insertion," *International Journal on Cryptography and Information Security* 3, no. 3 (2013): 23–30.

Roslin Nesa Kumari, G., Vijayakumar, B., Sumalatha, L., and Krishna, V. V., "Secure and Robust Watermarking on Grey Level Images," *International Journal of Advanced Science and Technology* 11 (2009).

Sadreazami, H., Ahmad, M. O., and Swamy, M. N. S., "Multiplicative Watermark Decoder in Contourlet Domain Using the Normal Inverse Gaussian Distribution," *IEEE Transactions on Multimedia* 18, no. 2 (2016): 196–207.

Salivahanan, S. and Arivazhagan, S., *Digital Circuits and Design*, 3E, Vikas, New Dehli, India (2007): 1–8.

Samuelson, P., "Copyright and Digital Libraries," *Communications of the ACM* 38, no. 4 (1995): 15–21.

Sánchez Alba, M. G., Ricardo Alvarez, G., and Sully Sánchez, G., "Architecture for Filtering Images Using Xilinx System Generator," *International Journal of Mathematics and Computers in Simulation* 1, no. 2 (2007): 101–107.

Sarkar, A., De, S., and Sarkar, C. K., *VLSI Design and EDA Tools*, 2E, Scitech Publications (India) Pvt. Ltd. (2014).

Sayood, K., *Introduction to Data Compression*, Morgan Kaufmann Publishers, San Francisco (1996).

Seo, Y. H. and Kim, D. W., "Real-Time Blind Watermarking Algorithm and its Hardware Implementation for Motion JPEG2000 Image Codec," in *Proceedings of the 1st Workshop on Embedded Systems for Real-Time Multimedia*, ACM, New York (2003): 88–93.

Shah, P., Meenpal, T., Sharma, A., Gupta, V., and Kotecha, A., "A DWT-SVD Based Digital Watermarking Technique for Copyright Protection," in *International Conference on Electrical, Electronics, Signals, Communication and Optimization*, Visakhapatnam, India (2015): 1–5.

Shaikh, S. and Deshmukh, M., "Modulation of Watermarking Using JND Parameter in DCT Domain," *International Journal of Engineering and Advanced Technology* 3, no. 2 (2013): 158–161.

Shi, J. and Malik, J., "Normalized Cuts and Image Segmentation," *IEEE Transactions on Pattern Analysis and Machine Intelligence* 22, no. 8 (2000): 888–905.

Sinha Roy, S., Basu, A., and Chattopadhyay, A., "Hardware Implementation of a Visual Image Watermarking Scheme Using Qubit/Quantum Computation through Reversible Methodology," in *Quantum-Inspired Intelligent Systems for Multimedia Data Analysis*, IGI Global, Ch. 4 (2018): 95–140.

Sinha Roy, S., Das, M., Basu, A., and Chattopadhyay, A., "FPGA Implementation of an Adaptive LSB Replacement Based Digital Watermarking Scheme," in *2018 International Conference (ISDCS)*, IEEE, Piscataway, NJ (2018).: 1–5.

Srinivas, N. and Deb, K., "Multiobjective Optimization Using Nondominated Sorting in Genetic Algorithms," *Evolutionary Computation* 2, no. 3 (1994): 221–248.

Stevens, G. W. W., *Microphotography and Photo Fabrication at Extreme Resolutions*, Chapman & Hall, London (1968).

Storn, R. and Price, K., "Differential Evolution – A Simple and Efficient Heuristic for Global Optimization over Continuous Spaces," *Journal of Global Optimization* 11, no. 4 (1997): 341–359.

Subramanian, R., Katti, H., Sebe, N., Kankanhalli, M., and Chua, T. S., "An Eye Fixation Database for Saliency Detection in Images," in *European Conference on Computer Vision*, Springer, Berlin, Germany (2010): 30–43.

Sur, A., Sagar, S. S., Pal, R., Mitra, P., and Mukherjee, J., "A New Image Watermarking Scheme Using Saliency Based Visual Attention Model," in *IEEE Annual India Conference (INDICON)*, IEEE, Piscataway, NJ (2009): 1–4.

Susanto, A., Setiadi, D. R. I. M., Sari, C. A., and Rachmawanto, E. H., "Hybrid Method Using HWT-DCT for Image Watermarking," in *2017 5th International Conference on Cyber and IT Service Management (CITSM)*, IEEE, Piscataway, NJ (2017): 160–170.

Tacticus, A., *How to Survive Under Siege / Aineias the Tactician*, Clarendon Ancient History Series, Clarendon Press, Oxford, UK (1990).

Tanaka, K., Nakamura, Y., and Matsui, K., "Embedding Secret Information into a Dithered Multilevel Image," in *Proceedings of the IEEE Military Communications Conference*, IEEE, Piscataway, NJ (1990): 216–220.

Tatler, B. W., "The Central Fixation Bias in Scene Viewing: Selecting an Optimal Viewing Position Independently of Motor Bases and Image Feature Distributions," *Journal of Vision* 7, no. 14 (2007): 1–17.

The USC-SIPI Image Database. Available: http://sipi.usc.edu/database/d atabase.php?volume=misc.

Tirkel, A. Z., Rankin, G. A., Van Schyndel, R. M., Ho, W. J., Mee, N. R. A., and Osborne, C. F., *Electronic Water Mark. Digital Image Computing: Techniques and Applications*, Macquarie University, Sydney, Australia (1993).

Tissandier, G., *Les Merveilles de la Photographie*, Librairie Hachette et CIE, Paris, France (1874).

Tiwari, R., Kaur, N., and Kaur, M., "An Optimization Image Watermarking Technique Using Biogeography Based Optimization," *International Journal of Engineering and Science (IJES)* 2, no. 3 (2013): 64–70.

Toet, A., "Computational versus Psychophysical Image Saliency: A Comparative Evaluation Study," *IEEE Transactions on Pattern Analysis and Machine Intelligence* 33, no. 11 (2011): 2131–2146.

Totla, R. V. and Bapat, K. S., "Comparative Analysis of Watermarking in Digital Images Using DCT and DWT," *International Journal of Scientific and Research Publications* 3, no. 2 (2013): 1–4.

Treisman, A. M. and Gelade, G., "A Feature Integration Theory of Attention," *Cognitive Psychology* 12, no. 1 (1980): 97–136.

Tsai, C., Chiang, H., Fan, K., and Chung, C., "Reversible Data Hiding and Lossless Reconstruction of Binary Images Using Pair-Wise Logical Computation Mechanism," *Pattern Recognition* 38, no. 11 (2005): 1993–2006.

Veeramani, S. and Rakesh, Y., "An Optimization Technique for Image Watermarking Scheme," *International Journal of Computer Trends and Technology (IJCTT)* 5, no. 3 (2013): 139–144.

Verma, M. and Yadav, P., "Capacity and Security Analysis of Watermark Image Truly Imperceptible," *International Journal of Advanced Research in Computer and Communication Engineering* 2, no. 7 (2013): 2913–2917.

Wagh, K. H., Dakhole, P. K., and Adhau, V. G., "Design & Implementation of JPEG2000 Encoder Using VHDL," in *Proceedings of the World Congress on Engineering*, Vol. 1, International Association of Engineers, London (2008): 670–675.

Wang, H., Ding, K., and Liao, C., "Chaotic Watermarking Scheme for Authentication of JPEG Images," in *IEEE International Symposium on Biometrics and Security Technologies*, IEEE, Piscataway, NJ (2008): 1–4.

Wang, Z., Bovik, A. C., Sheikh, H. R., and Simoncelli, E. P., "Image Quality Assessment: From Error Visibility to Structural Similarity," *IEEE Transactions on Image Processing: A Publication of the IEEE Signal Processing Society* 13, no. 4 (2004): 600–612.

Wilkins, J., *Mercury: Or the Secret and Swift Messenger: Shewing, How a Man May with Privacy and Speed Communicate His Thoughts to a Friend at Any Distance*, 2E, Rich Baldwin, London, UK (1694).

Willard, R., "ICE (Identification Coding, Embedded)," in *74th Convention of the AES*, Audio Engineering Society (1993): Preprint 3516 (D2–3).

Wille, R., Le, H. M., Dueck, G. W., and Grobe, D., "Quantified Synthesis of Reversible Logic," in *Design, Automation and Testing in Europe*, IEEE, Piscataway, NJ (2008): 1015–1020.

Wójtowicz, W. and Ogiela, M. R., "Digital Images Authentication Scheme Based on Bimodal Biometric Watermarking in an Independent Domain," *Journal of Visual Communication and Image Representation* 38 (2016): 1–10.

Wolfgang, R. B. and Delp, E. J., "A Watermarking Technique for Digital Imagery: Further Studies," in *Proceedings of the International Conference on Imaging Sciences, Systems, and Technology*, Las Vegas, NV (1997): 279–287.

Wong, M. L. D., Lau, S. I. J., Chong, N. S., and Sim, K. Y., "A Salient Region Watermarking Scheme for Digital Mammogram Authentication," *International Journal of Innovation, Management and Technology* 4, no. 2 (2013): 228–232.

Woo, C. I. and Lee, S.-D., "Digital Watermarking for Image Tamper Detection Using Block-Wise Technique," *International Journal of Smart Home* 7, no. 5 (2013): 115–124.

Wood, H., "Invisible Digital Watermarking in the Spatial and DCT Domain for Color Images," State College, Adams, Colorado (2007).

Wu, Q., Zhu, C., Li, J. J., Chang, C. C., and Wang, Z. H., "A Magic Cube Based Information Hiding Scheme of Large Payload," *Journal of Information Security and Applications* 26 (2016): 1–7.

Wujie, Z., Lu, Y., Zhongpeng, W., Mingwei, W., Ting, L., and Lihui, S., "Binocular Visual Characteristics Based Fragile Watermarking Scheme for Tamper Detection in Stereoscopic Images," *International Journal of Electronics and Communications* 70, no. 1 (2016): 77–84.

Xiang-yang, W., Yu-nan, L., Shuo, L., Hong-ying, Y., Pan-Pan, N., and Yan, Z., "A New Robust Digital Watermarking Using Local Polar Harmonic Transform," *Journal of Computers and Electrical Engineering* 46 (2015): 403–418.

Xu, H., Wang, J., and Kim, H. J., "Near-Optimal Solution to Pair Wise LSB Matching via an Immune Programming Strategy," *Information Sciences* 180, no. 8 (2010): 1201–1217.

Yang, C. H., "Inverted Pattern Approach to Improve Image Quality of Information Hiding by LSB Substitution," *Pattern Recognition* 41, no. 8 (2008): 2674–2683.

Yang, M.-S., "A Survey of Fuzzy Clustering," *Mathematical and Computer Modelling* 18, no. 11 (1993): 1–16.

Yip, S.-K., Au, O. C., Ho, C.-W., Wong, H.-M., and Li, R-y., "New Digital Watermarking for Few-Color Images," in *International Conference on Image Processing*, IEEE, Piscataway, NJ (2006): 2553–2556.

Zemcik, P., "Hardware Acceleration of Graphics and Imaging Algorithms Using FPGAs," in *Proceedings of the 18th Spring Conference on Computer Graphics*, ACM Press, New York (2002): 25–32.

Zhang, L., Tong, M. H., Marks, T. K., Shan, H., and Cottrell, G. W., "SUN: A Bayesian Framework for Saliency Using Natural Statistics," *Journal of Vision* 8, no. 7 (2008): 32.1–32.20.

Zhu, W., Xiong, Z., and Zhang, Y.-Q., "Multiresolution Watermarking for Images and Video," *IEEE Transactions on Circuits and Systems for Video Technology* 9, no. 4 (1999): 545–550.

Index